The Creative City

Cultural policies and urban regeneration between conservation and development

P.I.E. Peter Lang

Bruxelles · Bern · Berlin · Frankfurt am Main · New York · Oxford · Wien

Alessia Usai

The Creative City

Cultural policies and urban regeneration between conservation and development

ENCATC Book Series
Cultural Management and Cultural Policy Education
Vol. 2

The European Commission support for the production of this publication does not constitute an endorsement of the contents which reflects the views only of the authors, and the Commission cannot be held responsible for any use which may be made of the information contained therein.

Cover Illustration: The restored medieval village of Tratalias, Sardinia (IT) © Alessia Usai

Translations done by the International English Centre (Cagliari Office) and Traduzioni Web.

This publication has been peer-reviewed.

Éditions scientifiques internationales
Brussels, 2016
1 avenue Maurice, B-1050 Bruxelles, Belgique
info@peterlang.com; www.peterlang.com

Printed in Germany

ISSN 2466-7137
ISBN 978-2-8076-0192-5
ePDF 978-2-8076-0193-2
ePUB 978-2-8076-0194-9
MOBI 978-2-8076-0195-6
D/2016/5678/86

Bibliographic information published by "Die Deutsche Nationalbibliothek"

"Die Deutsche Nationalbibliothek" lists this publication in the "Deutsche Nationalbibliografie"; detailed bibliographic data is available in the Internet at http://dnb.d-nb.de.

Table of Contents

List of Figures 11

List of Tables 13

Acknowledgements 15

1. The Concept of Creativity 17
1.1 A critical review of literature on creativity 20
1.2 The urban dimension of creativity: different schools, different definitions 22
1.3 The link with spatial planning 23
1.4 The link with innovation 28
1.5 The link with culture, cultural policies and cultural programs ... 32

2. Strategies for Creativity 39
2.1 Creativity and the historical urban landscape read through the international charters and declarations 39
2.2 UNESCO programming for creativity and culture 46
2.3 European programming for creativity and culture 52
2.4 Conclusions 60

3. Creativity at the Regional and Urban Scale 61
3.1 The international context 61
3.2 The European context 63
3.3 The Italian context: the national dimension 67
3.3.1 Cultural and creative economy in Italy (1990-2009) 67
3.3.2 The creative capacity of culture in time of austerity (2010-2014) 76
3.4 The Italian context: the regional dimension 83
3.5 Conclusions 86

4. Regional Planning 89
4.1 The origins of the advanced cultural district: the industrial district 89
4.1.1 The classical district: a new economic-commercial approach in industrial organization (1920-1970) 89

4.1.2 The growth of the socio-technical dimension within the economic-commercial approach (1970-1990) 90
4.1.3 Industrial districts in the global market: a socio-cultural approach in industrial organization (1995-2005) 92
4.1.4 Advanced industrial districts: a cognitive approach in industrial organization and technological innovation (2005-2014) 96
4.2 From the industrial district to the advanced cultural district: the Italian contribution to the international debate 100

5. The Advanced Cultural District in Regional Planning 109
5.1 The existent instruments and their effectiveness 110
5.1.1 The economic survey: statistical indicators, analysis of the relationships between companies and production chains 110
5.1.2 The spatial survey: distribution (mapping) and context analysis 120
5.1.3 The sociological survey: the appraisal of competencies and resources 125
5.2 Building an advanced cultural district: methodological and instrumental aspects 127
5.2.1 The procedure 128
5.2.2 The feasibility study 133
5.3 Conclusions 135

6. *Creative Spatial Policies* for Cultural Heritage and Landscape 137
6.1 Assessment of the strategic positioning 139
6.1.1 Resource analysis 139
6.1.2 Relational analysis 142
6.1.3 Evaluation of the strategic-organisational positioning 143
6.2 The *integrated landscape project* 144
6.2.1 Definition of the project "Masterplan" 144
6.2.2 Identification of the governance model 147
6.3 Potential applications 156

7. A First Application 159
7.1 The plan as an instrument for the regulation of the relationships between territory, population and economic activities 161

7.2 Planning in Italy 163
7.3 Landscape planning in Sardinia 165
7.4 Integrated landscape programs established by the 2006 Regional Landscape Plan of Sardinia Region 168
7.5 A shared methodology for programs dedicated to cultural heritage and landscape in Sardinia 172
7.6 The proposed methodology: potentials, limits and future perspectives in research.. 174

Bibliography 175

List of Figures

Figure 1. Theoretical approaches in economic literature on creativity 21
Figure 2. Cultural heritage chain........ 45
Figure 3. The fields of action of UNESCO: organizations and activities from 2004 to 2014........ 51
Figure 4. MDG-F Portfolio: distribution of resources for activity..... 51
Figure 5. Culture in the 2014-2020 European Programming........ 58
Figure 6. Contribution of the cultural and creative field to the national economy in the different phases of the cultural chain 74
Figure 7. The creative district: a theoretical model 98
Figure 8. The advanced cultural district in Italy: a system of systems........ 107
Figure 9. The cluster of laser technologies used for the conservation of the cultural heritage in Florence 113
Figure 10. The matrix applied to the relational system of the Piceno University Consortium (IT)........ 114
Figure 11. Cluster analysis: the feasibility study for an advanced cultural district in the Province of Ascoli Piceno (IT)....... 117
Figure 12. Cluster analysis: "Reggio Emilia Città Creativa" project........ 118
Figure 13. Spatial distribution of creative companies with respect to the population density in Ljubljana (Slovenia) 121
Figure 14. Spatial distribution of creative companies in relationship to the year of construction of the building, with creative hotspots (black circles)........ 121
Figure 15. Design factors: the *Urban Network Analysis* 123
Figure 16. Procedural steps for the construction of advanced cultural districts starting from an *integrated landscape project* (masterplan)........ 138
Figure 17. The SWOT matrix 144
Figure 18. Strategic-organisational positioning matrix........ 144
Figure 19. Valorisation of cultural heritage in Italy: management forms according to the national laws........ 148

Figure 20. The sources of communication for the construction of an advanced cultural district 155
Figure 21. The strategies adopted by *place branding organisations* for the construction of an advanced cultural district 156

List of Tables

Table 1. Economic theories and evaluation of the performance in the cultural and creative industries (CCIs) 28
Table 2. Key concepts in the systemic approach 30
Table 3. Features of cultural policies and cultural planning 33
Table 4. Criteria for evaluation of the integrity of the asset 42
Table 5. Thematic priorities in Cohesion Policy 2014-2020 58
Table 6. Legislative interventions concerning the public assets of historical-cultural value 68
Table 7. The cultural and creative field in the KEA report (2006) 70
Table 8. Distribution of the interviews in the different phases of the cultural heritage chain 72
Table 9. The cultural and creative sector in the White Paper 73
Table 10. Strategic options for the development of the economies of scale and agglomeration in Italy 75
Table 11. The relationships between enterprises in the industrial district 93
Table 12. The new forms of district 99
Table 13. Industrial and cultural district compared 101
Table 14. Classification of cultural districts: economic-institutional features 103
Table 15. The components of the advanced cultural district: economic-institutional characteristics 108
Table 16. The three dimensions of creative spatial policies according to the eco-systemic approach 109
Table 17. International statistic frameworks 110
Table 18. Italian statistic frameworks 111
Table 19. Interdependence matrix 119
Table 20. Context factors: classification and definition in the *Genoa Creative Cities* project 122
Table 21. Management factors: classification and definition in the *Genoa Creative Cities* project 124
Table 22. Matrix of "internal consistency" for the *Creative Cities* project in Ljubljana (Slovenia) 125

Table 23. Example of an appraisal of competencies and resources for a cultural district 126
Table 24. Assessment criteria for the reports provided by the technical committee. 131
Table 25. Assessment system applied to the Cultural Districts in Lombardy and Veneto (IT): the development factors analysed 132
Table 26. Basic competences required to a *network manager* operating in the cultural and creative sector 141
Table 27. Analysis of the inter-organizational relations 142
Table 28. Logical Framework 146
Table 29. SWOT matrix for the assessment of the legal forms identified for the management of the future district 149
Table 30. The tools of negotiated planning to implement the interventions included in an *integrated landscape project* in Italy 153
Table 31. Framework for the analysis of the integrated landscape projects 160
Table 32. Participation tools in the landscape planning of Sardinia Region 166
Table 33. The Integrated landscape programs awarded and financed by Sardinia Region 171
Table 34. Comparison between the two methodological approaches: application of the framework 173

Acknowledgements

The present work derives from a positive path of research and personal growth which would not have been possible without the guidance of my PhD Supervisor, Professor Anna Maria Colavitti. To her goes my special thanks for supporting me along my PhD Program and along the experience of the 2015 *ENCATC Research Award.*

I express my gratitude to the *European Network of Cultural Administration Training Centers* (ENCATC), to Annick Schramme – the ENCATC President –, to the Scientific Committee, to the Members of the Jury of the *ENCATC Research Award,* and to Peter Lang Publishing Company to give international breath to my research and make the publication of this volume possible.

In addition to this, my gratitude goes to Sardinia Region which supported my research with the resources of P.O.R. SARDEGNA F.S.E. 2007-2013 – Objective regional competitiveness and occupation, Line IV - Human Resources, Activity l.3.1 "Funding of PhD courses aimed at the training of highly specialized human resources, in particular for the ICT sectors, nanotechnologies and biotechnologies, energy and sustainable development, agri-food industry and traditional materials."

A special acknowledgment goes to the professors and researchers of the Department of Civil Engineering, Environmental Engineering, and Architecture (DICAAR) and the Academic Board of the PhD Programme in Technologies for the Conservation of Architectural and Environmental Heritage of the University of Cagliari (IT), who helped me to improve my thesis in its critical aspects. Furthermore, my gratitude goes to Professor Nicholas Clifton, my mentor during my research experience at the *Cardiff Metropolitan University.*

Lastly, a special thought to my family for supporting me every day.

The Concept of Creativity

Spatial Planning, Innovation and Culture

In the last thirty years, several disciplines have tried to define creativity in order to provide a solid foundation to economic theories inspired by this concept. In psychometrics and in cognitive psychology, creativity has been conceived as a set of acts, ideas or products which modify an already existing domain or generate a new one (Csikszentmihalyi, 1988, 1996). This definition introduced two relevant concepts for the economic and entrepreneurial world. Ideas, acts, and products need a circumscribed reference framework, like a market, and within it a social group creating demand and recognizing them as such in order to be created. From this point of view, the 'creatives' generally belong to that group of people who are able to recognize innovative ideas present in a certain domain, can make them theirs, and manage to cultivate them with little effort in order to re-sell them at a high price when a client recognizes their potential.

In the analysis of the reference framework different features of creativity (artistic, scientific, economic, and technologic) have been individuated (Amabile, 1983, 1998). At the same time, Torrence, Sternberg, and Lubart (1991) individuated six different elements which affect the creative's choices. These factors are: personal predisposition (ability of analysis, synthesis, and manual abilities); the degree of knowledge of the domain; personal thought; motivation; personality; operational environment. The latter is particularly important to Sternberg (2006), who considers the context as a necessary and crucial element in the reading of the reasons and the strategies underlying the operational choices of the creatives. In fact, he states that a taxonomy of creativity based only on reference domains cannot be considered sufficient. Nonetheless, it is only in Staber (2011) that the context becomes a distinguishing feature for a new classification of the studies on creativity according to three different surveying approaches.

The first, defined *character-based* or *cognitive*, focuses on the individual abilities of the creatives: highly qualified and professionally independent people whose identity is linked to the success of their ideas and, in particular, to the way they are perceived. An idea that was successful in the past is rarely modified. Changes either happen due to mistakes made by other subjects during its reproduction, or because of a choice made by the creator under the influence of other elements (people or ideas). This is a representation linked to the romantic idea of geniuses or artists working in solitude, which are far away from the real and current conditions of training and working that contemporary creative talents have (Lai, 2006; Larsson, 2001).

The second approach, called *relational*, focuses on the immediate relationship built inside a specific environment to acquire information, material resources and to find similarities with others (*creative networks*). This is based on the assumption that no individual is completely isolated in economic life, and that inspiration derives from interaction and reciprocal support. This approach can lead to a distorted view of the existing economic exchanges, which are naturally characterized by a certain degree of competition. The environment is considered the spatial support of the dynamic relationships of cooperation/competition among the creatives.

Lastly, the *systemic* approach concentrates on the interdependence between groups of individuals and creative systems (*clusters*) and their operational system[1]. One of the advantages of the systemic approach is that it calls attention to the interdependence of the different fields (e.g. physical, social, cultural, technological) and the fact that the boundaries of their related spatial systems are constantly changing.

The relational and the systemic approach strongly emphasize the collective dimension of creativity and therefore seem linked to the concepts of *strategy* and *habitus* elaborated by Barth (1981), Bourdieu and Wacquant (1992) in the field of sociology and urban anthropology. Before these authors, social sciences were characterized by the contrast of two schools of thought: determinism, associated to Adam Smith, and structuralism, associated to Emile Durkheim (Elster, 1989). Determinism, also called methodological individualism, motivates every individual action within the social sphere with the attainment of the best economic advantage (*instrumental rationality*). This is according to the behavior model of the *homo oeconomicus.*

Structuralism, on the other hand, considers individual actions as the result of the conditioning imposed by the social and cultural structures. Here the word "structure" means both a stable social organization (Levi-Strauss, 1983) and a *long-lasting* connection or attitude toward a certain environment which time cannot erode and which is repeated for infinite generations. According to the behavioral model of the *homo sociologicus* (Braudel, 1969).

The resolution of the conflict between freedom and determinism, between human agent and structure, derives from the introduction of the be-

1 The systemic approach is also adopted in the studies on cultural *clusters* and regional innovation systems (Asheim and Isaksen, 1997; Cooke and Lazzeretti, 2008).

havior model of the *homo strategicus*, according to the research of Pierre Bourdieu and Friedrik Barth. According to it, the individual has an active role, if compared to social and cultural norms, because he makes them his own when these are in favor of him, while he tries to avoid them when they damage him, by adopting a strategic approach. This design includes:

- The notion of *habitus* or "socialized subjectivity" by Bourdieu (1985), which explains how individuals incorporate schemes of action and particular values to adapt their contest or to a new one, schemes dictated by "practical sense" or "sense of play," rather than economic profit;
- the notion of "choice" and "strategy of the players" by Barth (1981), which concentrate on the individuals and their choices, *cumulatively* capable of modifying the same social structure.

According to Wacquant (1992) and Davis (1994), Bourdieu and Barth's theories solve the problem of continuous "invention of cultures", historical culture as well, though the principle that social convention is continuously reinvented during the action, according to processes of individual and social creativity. This is perhaps the reason why sociology and urban anthropology have been the first disciplines which tried to understand and evaluate the *input* and the *output* of 'creative systems' (institutional centres, creative fields, districts, *network, clusters*, etc.); analyze the relations between the various components of the system and between systems themselves (local vs. global, physical proximity vs. relational proximity, cooperation vs. competition, etc.) (Cooke and Lazzeretti, 2008; Marshall, 1920; Beltram Niessen, 2007; Santagata, 2006; Scott, 2006).

Recent studies on ecology marked a step forward in research on creativity, moving the attention from its spring, man, to the subject, ideas. According to this new "evolutionistic" approach, a "population" of ideas all equally able to capture human attention – and therefore in continuous competition – coexists within the reference framework, i.e. the market.

The evolution of the idea is determined by the unity of selections (creative systems) which establish criteria of environmental selection (elements of innovation in ideas) and the environmental load (how many ideas can receive attention). Therefore, the population of ideas can grow until it reaches the environmental load or until the criteria of environmental selection do not change, rewarding different ideas. When these events take place, some ideas survive, while the others are forgotten and substituted with new ideas which better suit circumstances (Staber, 2011). It is an open model where all creatives find space both as individuals and as group/system, defining criteria of selection of the ideas, in the first place, and their environmental load, in the second place. In any case, the consequences of the action of the creatives on the reference market-ecosystem are deep.

1.1 A critical review of literature on creativity

The different disciplinary interpretations of creativity provide some useful hints to accomplish a critical revision of literature on creative economy. According to the entrepreneurial definition of cognitive psychology it is possible, for instance, to recognize a group of studies characterized by a *people-based* approach, in where "the work follows the people". Creative workers are here described as highly qualified workers and independent people who chose to settle where the quality of the surrounding environment (*soft infrastructure*) allow them to best develop their ideas, independently from the presence of big companies in the field where they operate. This is the research branch where the theory of *Creative Class* and the model of 3Ts were successfully created by Richard Florida (1995, 2002, 2005)[2].

Of course, the focus on the better prepared and highly-performing component of productive society opens this approach to some criticism, such as the expansion of the social inequality and segregation[3] (Florida and Tinagli, 2004, 2005; Jacobs, 1961; Markusen, 2006; Zukin, 2008, 2010). Another critical aspect is the generality of the postulates on which it is bases, including the concept of creative 'class,' which makes the definition of appropriate policies for the cultivation and the attraction of talents in a landscape marked by the absence of long-term policies on a national and regional level (Markusen, 2006; Clifton, 2008).

In opposition to the theory of creative class, other studies support the importance of *hard infrastructures* and the geographic factor in the life choices of the creatives (proximity[4], economy of agglomeration).

At root of this relational approach there is the idea that workers keep settling where the market offers better working conditions, where large companies are established or, even better, large industrial areas embody different economies of scale. The studies about creative and cultural industries, cultural institutions, innovation and knowledge centers, and their agglomerations (*network*, *cluster* and district) fit in this research branch

---

2 See the statistic reports based on the *Global Creative Index*, on the *European Creative Index* and on the *Italian Creative Index*.

3 In spite of the *Global Creativity Index* and the *European Creativity Index* recognizing the *cluster* of Scandinavian Countries as a sustainable model of creative economy for a long-term development (Florida and Tinagli, 2004; Florida *et al.*, 2011)

4 In this sense, the studies related to the concept of proximity are important. They highlight how the location advantage turns into a limit for a creative system if the relationships between its components become too narrow or, on the contrary, to weak (Boschma, 2005).

(Marshall, 1920; Becattini, 2004; Santagata, 2006; Scott, 2006; Beltram Niessen, 2007; Cooke and Lazzeretti, 2008).

Similarly to what happened on an epistemological level with the contribution of ecology, the work of John Howkins (2001, 2010) moved the attention from the creative individuals to the environment in which they operate and to the ideas produced within it:

> We need to rethink the reality of creativity and the creative economy. We need new coordinates and a new compass. The new coordinates must take account of the environment in which people have ideas in their ordinary (or extraordinary) lives. I call this the "creative ecology". An effective ecology provides the habitat where an organism can live and flourish. A creative ecology is one where all people have the freedom to pursue their own ideas and where markets allow them to exchange ideas. Its main characteristics are soft infrastructure, social networks and social markets. We are developing parameters to measure this ecology (Howkins, 2010).

A new phase is foreseen for the research. Creativity is taking a comprehensive dimension which looks at society, economy and territory as a whole, without addressing only a cultural and creative 'class,' 'industry,' or a 'neighborhood/district' (Staber, 2011) (see Figure 1).

Figure 1. *Theoretical approaches in economic literature on creativity*

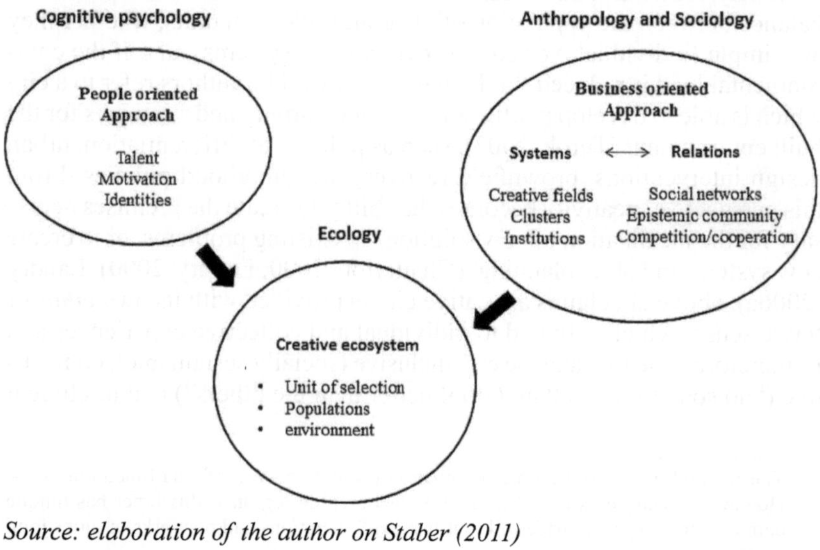

Source: elaboration of the author on Staber (2011)

1.2 The urban dimension of creativity: different schools, different definitions

The application of the aforementioned theoretical approaches to the urban dimension leads to different definitions of creative city[5]. According to Richard Florida's theory, a city is creative if it is able to adopt strategies for the cultivation and attraction of the talents through: 1) school and occupational policies that increase the *skills* of the resident community; 2) policies of cooperation, exchange and immigration which ease the input and settlement of highly qualified workers coming from other contexts (regional or national). On the contrary, in a business-oriented approach, a city is considered creative when it hosts, grows and supports the creative systems of the territory, whether they are made by large industrial companies, public institutions or productive districts. In case of *clusters* or *networks*, a creative city has to guarantee its support in the handling of long distance relationships and at the same time provide adequate solutions to the networks or the external groups which decide to move to the city. The most widespread instruments are policies for research and innovation, and for the definition/realization of specific urban spaces accompanied by particular forms of economic promotion (e.g. Pharmaceutics districts, Cities of Science and Research, Centers for heritage conservation, artistic quarters, etc.).

Lastly, Howkins and Staber consider a city as 'creative' if it presents a balance between the population of ideas and selection units, whether they are simple (individuals) or complex (relational systems), and if the environmental load is reduced (high quality of life). The authors refer to a city which is able to develop public policies for creativity and strategies for the built environment (Turok, 2007), such as policies of differentiation, urban design interventions, brownfield recovery, neighborhood policies. From this perspective creativity becomes the ability to create the premises necessary for the idealization of new solutions to existing problems, or to create new systems in public planning[6] (Chatterton, 2000; Landry, 2000). Landry (2006a), above all, claims a creative city is provided with its own *ethos* or 'civic sense' which is linked to individual and collective experiences and is, therefore, able to make the city inclusive (socially sustainable), competitive ("do something well and do it better than the others") and intelligent

5 The research is focused on metropolitan cities and network of small-medium cities. However, it can be extended to a whole geographic region if this latter has unique complex urban system made of big urban agglomerations and networks of cities joint together.

6 This research path is theoretically and empirically autonomous, making it completely separated. It finds its roots in the work of Michel De Certeau (1990) and Hannerz (1992) and in a historical perspective in Hall (2000) (Beltram Niessen, 2007).

(learning from its own mistakes). The creative *ethos* of a city is revealed through the political and institutional framework (professionalism and efficiency), local identity (distinctive, diversity, local buzz, expressiveness, resilience), social environment (opening, credibility, accessibility, and participation), the local productive system (entrepreneurship, research, and innovation), ruling class (strategic leadership, flexibility, immigration), training (talent development, learning systems), communication (connectivity and networking), quality of places (perceived and realized) and quality of life (livability, wealth of the population) (Bathelt *et al.*, 2004).

1.3 The link with spatial planning

Within the discourse on the creative city, urban planning finds its space with respect to the location choices of the creative systems and in relation to the definition of a urban environment suitable for the *creative class*. This emerges both in Anglo-Saxon literature and in the Mediterranean one. Where creativity is often associated to urban regeneration and local development[7] (Andersson *et al.*, 2011; Bianchini, 1990; Evans, 2001; Grogan and Mercer, 1995; Landry, 2006b).

Nonetheless, in everyday life, the interventions seem to occur in an almost self-referential manner: iconic buildings, mega-campuses, creative neighborhoods, programs linked to the theme of requalification, both generic and external to the planning which, once the planning cycle is finished, degrade or, on the contrary, trigger processes of gentrification which damage the community of residents (Evans, 2001; Turok, 2007).

The reason is to be found in the recent change of economic paradigms with the affirmation of a *new public management* and with a shift of competences from governments to partially private institutions[8] (Sager, 2011). The consequences for urban and regional planning have not been fully analyzed, but the recent economic crisis and the cuts to government spending proved how neo-liberal planning are path-dependent and context-specific (Peck *et al.*, 2009). Creative spatial policies do not escape this logic, especially those linked to cultural heritage, where the presence or absence of a strong public-control tradition is fundamental for the development of a culture-centered vision of creativity (Andres and Chapain, 2012; Tabellini, 2005). Limiting the discourse to neo-liberal planning, creative spatial

7 In the Italian context, see Carta (2009) for the studies on the *Creative Class* (in which the T of territory is added to the canonical 3Ts by Florida), Camagni (2008) for the concept of territorial capital and Casavola and Trigilia (2010) for new models to assess local development.

8 In British literature, these entities are defined as *Quasi-Autonomous Non-Governmental Organization*, in short *quangos*. They are organizations financed from tax payers but which are not directly controlled by the central government (Wettenhall, 1981).

practices are closely related with economic and local development and, on this basis, they can be classified in (Sager, 2011):

1. *Territorial Marketing*: the creation of territorial brands according to three groups of potential buyers (residents, companies, tourists).
2. *Urban development by appealing to the creative class*: it implies the creation of spaces for the consumption of goods and cultural activities, spare time, and residence, such as amusement and entertainment centers, artistic neighborhoods, gated communities, but also big events and nightlife.
3. *Economic Incentives*: they include investments of large companies o important financial transactions, (single brand) private business districts, urban free zones, recovery of disused areas, private managing of public space in urban commercial areas.
4. *Competitive offer*: the idea is to create half-private markets in public services pushing local organizations to compete in the allocation of resources and to improve the quality of organizational management. In this case, partnerships are an important condition.

From an institutional point of view, a difficulty emerges in governing and harmonizing the two souls of the new neoliberal planning, the business-climate one and the people-climate, within the relinquished processes the new public management implied (Clifton, 2011). In territorial marketing, for instance, the dichotomy between the image built for internal targets (residential, local businesses) and the one built for external targets (tourists, foreign investors – among which are the creative systems and creative class) questions place branding as a mere selling of the 'territory'[9]. This leads back to Pratt's critique (2008) to Richard Florida concerning a consumerist, market-driven vision of creativity, which culminates in the production of iconic buildings. The complete opposite of the Anglo-Saxon model of Charles Landry and of the *Department for Culture Media and Sport* (DCMS) who analyzed creativity in terms of process and production to amplify its political effect (creative *ethos*) and economical effect

9 For this reason place marketing is now focusing on the process of competitive identity building, i.e. the creation by institutional organizations of the field (e.g. Local Touristic Systems) of a credible reputation for a determinate place through: (1) a strategy for promotion and marketing which puts together the needs and the expectations of the actors involved (institutional, economic, social) and guides them toward their goals, clear and realistic at the same time; (2) acts and interventions which give *substance* to the prepared strategy, making it independent from political and institutional changes; (3) the repetition of highly expressive *symbolic acts* which ease the affirmation of the new reputation and imagine of the place, preventing the strategy to remain anonymous (Anholt, 2011).

(cultural and creative industries)[10]. Landry himself (2006b) recognizes how the two approaches are independent for chronological and circumstantial reasons. In fact he traces the debate on the creative city back to the early 1980s, associating it to the effort of the British, American, and Australian artistic community to show the contribution that art has provided to the economies of the *urbs* (‘*planning for the arts*’). In the following years, the author claims, the regenerative power of arts has been extended to social policies. Cultural resources and cultural planning, as defined by Franco Bianchini (1990) and Colin Mercer (1996), became the tools to improve equity and social potential in terms of imaginative power and problem-solving of the *civitas*. Vice versa the social creativity became the way to enhance cultural resources. When applied to the strongest fields of cultural production (cinema, publishing, computer science, etc.), creativity moved the research from fine arts to cultural industries, defined as “creative” industries by DCMS in 1997. Since 2000, Landry states, the theme of creative industries arrived in the European scene, according to theoretical models proposed by Anglo-Saxon literature. In this phase Richard Florida developed his theory. He concentrates in the role of individual talents (*creative class*), innovation of productive and territorial systems (cities, regions, states). Yet, he completely neglected the connection between art and urban regeneration of the previous discourse.

To quote Landry (2006b):

> The USA which had been so influential in getting the idea of the economic impact of the arts off the agenda had been very slow in seeing the link between the creative industries and the creative city. At the time writing his book [Richard Florida] had not been aware of the creative city debate.

What follows is that as charming and intuitive as it is, Florida’s theory alone cannot define the components and features of the creative city, which help the creation of a model for the planning of spatial policies integrated in the process of planning. For this reason, in the European and Mediterranean landscape policymakers, researchers and planners are working for an ecosystemic vision where the creative city by Charles Landry, the creative industries by the DCMS and the Creative class by Richard Florida are assessed and planned together through integrated statistic, economic and spatial tools[11].

---

10 However, Andy Pratt recognizes the limits within the cultural and creative industry. In particular, he underlies how Porter’s *cluster* is not efficient in the esteem of cultural and creative production coming from multinational businesses, which prevails over PMI and *Creative Artists* (*C-Art*) (Pratt, 2004). He forwards the *Actor Network Theory* (*ANT*) and the *creative value chain* as possible solutions (Pratt, 2000; Pratt and Taylor, 2006).

11 The recent studies on cultural-cognitive capitalism by Allen J. Scott (2012, 2014) are moving this way, as well as O’Connor and Shaw’s studies on social and civil value of the creative city (2014).

Among the most important analytic models developed according to this vision, it is important to remember the model elaborated by INTELI (2011) and Miguel Rivas (2011) within the URBACT project *Creative clusters in low density urban areas*, which introduced the concept of 'local creative ecosystem':

> An environment of excellence based on creative assets that generates socio-economic growth and development, and comprises three interlinked components: economy – creative industries, place – creative spaces, and people – creative talents (INTELI, 2011).

The model developed by Maurizio Carta (2007), which re-elaborates Florida's theory, is also significant, since it adds to Technology, Talent, and Tolerance, the T of Territory[12]. At the same time, Richard Florida was involved in the *Creative Cities Leadership Project* – CCLP[13] (2006-2008) and the *Knight Creative Communities Initiative – KCCI*[14] (2007-2008), via the *Creative Class Group*, reaching a similar result. He created a new 4T model where the third T means territorial assets[15], which is considered the means to involve the residents in the construction of creative communities. The creative districts elaborated on these bases depend on free mutual relations and they present the following basic features: spatial proximity, clustering, and other forms of closeness between creative and cultural industries (Boschma, 2005), the availability of common spaces of productions and forms of mixed residence (residence/job) for creative talents (*creative neighborhoods*) (Montgomery, 2003; Lai, 2006).

Keeping into account the aforementioned classification, an ecosystem-based approach allows to circumscribe the object of creative spatial policies to the cultural resources classified by Bianchini in the early 1990s[16].

They include:

1. Historical, artistic, archaeological, and anthropologic heritage;
2. Physical environment, including the architectural heritage, cultural landscapes and historic urban landscapes, topography of the territory;
3. Quality of public spaces;

---

12 Territory was added as a catalyst and was declined in the following factors (the "three Cs"): Culture, Competition, Cooperation (Colavitti, 2013).

13 The project involves the communities of El Paso, Texas and Tacoma, Washington (Stern and Seifert, 2008; Voelpel, 2007).

14 The project involves the communities of Charlotte, North Carolina and Duluth, Minnesota/Superior, Wisconsin, and Tallahassee, Florida (Stern and Seifert, 2008).

15 See: http://www.creativeclass.com/services/creative_communities

16 According to a "anthropological" idea of culture on which it will come back later.

4. Diversity in the spaces and facilities for leisure activities, entertainment and culture;
5. Visual arts, entertainment and cultural industries;
6. Local events promoted by charities and associations (such as carnivals, fairs, festivals, etc.);
7. Hobbies of the residents;
8. Culture of young people, ethnic minorities and other communities of interest present on the field;
9. The internal and external image of the territory, which is expressed through its various cultural representations (songs, myths, tourist guides, press reportages, radio and TV, internet, etc.);
10. The repertoire of products and local productive skills in manufacture, industry, and services.

The list helps to define the content of the creative spatial policies which often coincides with local cultural planning, meant as:

> Planning is the organizational foundation from which all other functions flow. As such it is much too important to be left in the hands of planners as we know them.
>
> Planners make spaces: people and communities in their daily activities make places come what may, and often at variance with the planners' original intentions. This is not a populist line: you still need the planners but, most importantly, you need to be able to broaden their agenda, to give them an ethical corrective in their designs and plans; an ethical corrective based on consultation and research rather than on the drawing board aesthetics of the utopian space and the masterplan (Mercer, 1996).

This image coincides only partially with the one adopted by Landry in the constitution of his *creative city index*[17] and it confirms that the ecosystem-based approach is now acquiring its own disciplinary autonomy.

---

[17] The method adopted by Charles Landry in the construction of his evaluation model is based on the following domains (www.charleslandry.com): (1) Political and institutional framework; (2) Distinctiveness, diversity, liveliness, and expressiveness; (3) Opening, reputation-credibility, acceptability, participation; (4) Entrepreneurship, exploration, and innovation; (5) Strategic leadership, agility, and vision; (6) Development of the talent and of the educational system; (7) Communication, connectivity, and networking; (8) Urban quality and urban design; (9) Livability and wellness; (10) Professionalism and efficiency.

1.4 The link with innovation[18]

Cultural and creative industries (CCI) produce goods and services on a global scale. The performance of their production processes are analyzed according to the economic theories on technological organization[19] - even though CCIs present some peculiarities compare to traditional industrial organization (see Table 1).

Table 1. Economic theories and evaluation of the performance in the cultural and creative industries (CCIs)

Theoretic approach	Evaluation models for cultural and creative industries
Neoclassic	Analysis of the production chain
Evolutionary	Analysis of inter-industrial relationships (input/output)
Systemic	Analysis of location according to Porter's *cluster*
Integrated	Analysis of the environment

Source: elaboration of the author

Firstly, the neoclassical theory focuses on the analysis of the allocation of scarce resources to maximize their profit, once the method of production has been evaluated. A fundamental feature of this model is the connection between production and distribution. Here the basic principle is to increase the offer of productive factors in order to obtain an income and to promote growth. Innovation is strictly linked to the tangible aspects of production (tools, means, procedures) while the growth of the social capital is considered an external factor. The only elements which impact the growth are workforce and initial capital (Giordani and Zamparelli, 2007).

In the 1970s and in the 1980s the evolutionary theory re-elaborates the concept of growth linking it to the technological change. Joseph Schumpeter (1939), pioneer of the studies on innovation, defines technological change as the set of organizational, behavior-related and relational changes enact by

18 The paragraph has been written at the end of a period of research and study at the Metropolitan University of Cardiff in 2013 under the scientific responsibility of Prof. Nicholas Clifton. This experience represented an important occasion of the investigation on analytic models applied to the analysis and evaluation of cultural and creative industries in the United Kingdom, in relation to regional planning, innovative systems and touristic and territorial marketing.

19 See United Nations (2011), pp. 77-82.

the players of a system. The process of innovation is described through the rigorous sequence of three phases[20] (Schumpeter, 1939):

1. invention: the production of new knowledge;
2. innovation: the application of the existing knowledge in the productive field;
3. diffusion: the diffuse usage of new technologies;

While innovation is mainly influenced by the location factor (and more generally by proximity), invention and diffusion are influenced by contextual mechanisms, meaning by the spatial changes produced by formal and informal social networks on a local level (Boschma, 2005; Iammarino, 2004). The concept of innovation is therefore extended from the research of new productive processes and/or products (*technical innovation*) to new organizational forms, which are able to open new markets (*technological innovation*).

During the 1990s, the systemic theory[21] expands the concepts of innovation to all the phases of the "Schumpeterian Trilogy". Starting from the idea that particular innovative activities can be both cause and effect, consequence and prerequisite of a technological change, the systemic theory concentrates on collective and interactive learning. The aim is to market new knowledge instead of products, production systems and organizational models ('experiential economy'). In the systemic model the phases of invention, innovation and diffusion are linked by a two-way relation and the key concepts are knowledge, learning[22], interaction, network (Table 2). Therefore, the innovative system becomes a set of elements and relationships interacting during the production, diffusion and use of some knowledge, new and profitable[23].

20 The so-called "Schumpeterian Trilogy," described in Robertson (1967) and Mahdjoubi (1997) and criticized for the rigidity of its waterfall or *trickle-down* model (Cooke, P. and Memedovic, 2003).

21 Drawn up by the neo-Schumpeterian school founded by Lundvall (1992) and Freeman (1987).

22 There are several forms of learning: by doing, by using, by interacting. In the systemic approach, the first two concern the innovation while the last one concerns distribution (Cooke, P. and Memedovic, 2003).

23 Belussi and Staber (2011) remind us that inventions become innovations only if they have a practical and economic outcome. They propose the outstanding case of Leonardo da Vinci for whom, in fact, only some inventions became reality.

Table 2. Key concepts in the systemic approach

Knowledge	**Learning**
Skills and abilities embedded in the routines and conventions of innovative firms and support organizations. Old and discarded or forgotten skills and abilities connected to traditional industrial productions	Collective process shaped by the existing structure of production, by organizations and by institutions
Interaction	**Network**
Regular means of formal and informal shared communication between relevant network organizations and their members aimed at learning, critiquing and creating projects and ideas.	An interconnected group or system put together to pursue common interests (in relation to an innovative project, product or process).

Source: elaboration of the author on Cooke (2001), Cooke e Memedovic (2003)

From the second half of 2000s, the systemic model evolves towards an integrated approach thanks to the studies on innovative regional systems. The new approach emphasizes the mechanisms acting *supra* and *infra* the innovation process, such as routine, technological trajectories, environmental selection, diversity and path-dependency. The requirement for technological change are institutional learning and interdependence between the structural components and the players of the innovation system. They should be considered a dynamic feedback mechanism: environmental features influence the players and vice versa the players influence the environment (Belussi, 2011; Iammarino, 2004). The analogy with Howkins' creative ecology is self-evident. In this theory the carrying ability of the environment constitutes a limit for the population of ideas on which the units of selection (individuals and creative systems) can act and at the same time the units of selection establish environmental criteria (elements of innovation) for which only some ideas survive.

The real connection between innovation and creativity, nonetheless, lies in the systemic definition of innovation of the new-Schumpeterian school. In short, it concerns the production of new knowledge through a process of collective learning linked to the construction/use of new technologies in the production (learning by doing, by using) or to the interaction with normal production activity, with other innovative systems or with the final customers (learning by interacting). Considering the definition by Csikszentmihalyi and setting as reference domain the technological one, innovation and creativity are both referred to a set of acts, ideas or products, which change an already existing domain, or generate a new one. Furthermore, creativity, as well as invention, presents an integrated and complex dimension linked to the knowledge elaborated and incorporated by cultural or aesthetic reference frameworks (values, symbols, handcrafted objects, and beliefs)[24]. For this reason, innovation and creativity are often considered synonyms. On the contrary, they actually present quite a strong difference in the purpose of their two processes: while creativity produces ideas which can also have a not immediate or utilitarian practical outcome, the invention in innovation process necessarily requires useful ideas for the economy of the production (Belussi, 2011, p. 7).

It derives that:

- the application of an innovation does not necessarily lead to a new product, since it can concern 'technical' aspects of production or organization;
- creativity represents the first step towards innovation because it provides ideas to the phase of invention. The opposite is not valid, since innovation means neither new products nor use of creativity or creative talents in production.

According to these considerations, it is possible to update Csikszentmihalyi (1996, 1998) defining:

- *creativity* as the set of acts, ideas or products which originates a new domain;
- *innovation* as the set of acts, ideas or products which introduces elements of novelty in a domain, in a production process or phase (Scott, 2010, p. 119).

The difference between creativity and innovation can be verified through the organizational forms and products of the cultural and creative industries which are more volatile, flexible and 'experiential' compare to the traditional ones:

> Creative products, thus, differ in their survival chances, depending on their market value but also in response to social and cultural conventions, such as those contained in fads and fashions. Related to the organizational approach, [...] Creativity is enacted differently in various settings.
>
> The project-based organization or the spontaneous networks built by artists represent the central organizational form in the artistic area; collaborative research networks are the locus of the inventive activity in science; and team-work, and communities of practices, characterize the setting of firms in manufacturing industries (Belussi, 2011, p. 26).

24 The systemic features of creative ecology are: (1) the context, for the evolution of the communities of players; (2) the tendency to gather and geographical proximity; (3) the transitory nature of cultural production (Staber, 2011).

Likewise to what we observed for creative spatial policies, the city represents a promising *habitat* for the processes of innovation, since it facilitates the monetization of potentially pre-existing agglomeration economies for economic operators (Casoni, 2011; Lai, 2006).

Within literature on innovation it is possible to individuate two general models of city: the *hub* and the *relational network*. The first one is founded on the advantages of urbanization economies: the possibility of personal contacts, availability of complementary goods and advanced services, dynamic coordination of economic players. On the contrary, the second one is founded on non-economic relationships which, thanks to spatial proximity, produce an exchange of information and knowledge, whether implicit or explicit. Implicit knowledge, in particular, concerns "everything that is accumulated in people's heads when they become expert in solving certain problems"[25] (Casoni, 2011). This highlights the importance of "face to face" relationships and of the "spirit of giving" in the technological process, establishing a further connection with the theories on creativity, according to which only the city with an appealing social environment can attract global fluxes[26]. This is in reference to the processes of creolization which are taking place in the American and European capital cities. These phenomena have now left the outskirts of the cities and are now extending to the metropolitan regions, the "urban constellations" and to the networks of medium-small cities (Castells, 2001; Lai, 2006).

1.5 The link with culture, cultural policies and cultural programs

In order to understand the connection between culture and creativity it is necessary to start from the concept of 'culture', its evolution in the last fifty years and its engagement in cultural policies and cultural planning (see Table 3).

Cultural policies are part the public policies of national States, which put them in practice through their Ministries/Departments or through local administrations/public organizations.

25 In the original text: "*tutto quello che realmente si accumula nella testa delle persone quando diventano esperte nel risolvere determinati problemi*" (Casoni, 2011).

26 Hall (1998) individuates a connection between creative processes and the following urban elements: remarkable demographic dimensions, considerable wealth, and ability of attracting creative talents.

Table 3. Features of cultural policies and cultural planning

	Cultural Policies	**Cultural Planning**
Perspectives	Based on *discipline.* A fragmented perspective driven by disciplinary "compartments" (theatre, dance, museums, etc.).	Based on *place.* A multidisciplinary perspective (better "understanding" of the origins and customs of the place).
Concept of Culture	Based on art. *Fine arts* and products of cultural industry of the place.	Based on *Cultural resources.* Cultural and natural assets of the place (the same usually considered in the studies on the local "heritage").
Logic of local intervention	*Business-oriented* attitude. "Art for art's sake"	Strategic logic. Emphasis on contributions to urban development
Decision-making process	Traditional role of the public administration: provider of funds and tools.	New role perspectives, more focused on the function of intermediation, which improve the social action of the players interested in the cultural dimension (on any level)

Fonte: Porello (2006)

Cultural policies developed during the 18th and 19th century in Europe, along with the birth of National States. They had an educational purpose connected to the National identity (preservation of cultural heritage as collective memory, strengthening of the feeling of citizenship). Their approach was culture-driven and based on the preservation and valorization of the historical-artistic heritage, even with new forms of organization[27]. They had a political connotation that in Anglo-Saxon and Italian literature is often associated to the social ideology of the historical left ("culture for everyone", "art for art's sake")[28].

On the other hand, cultural planning is shaped as a strategy that is able to reconcile different elements of local cultures within one field of action, which aims at local development according to a holistic and shared approach. It essentially deals with the creation of the management system and the regulations for the cultural assets within the framework of local

27 This is the case of the Italian cultural districts.

28 This is one of the reasons for which the saying 'cultural industry' has been substituted by 'creative industry' in the documents of the *Department for Culture Media and Sport* (DCMS) in England.

resources to produce positive economic and socio-spatial returns, beyond the planning.

Cultural planning born at the end of the 20th century in the United States, Australia, and Canada with the purpose of reaching cultural communitarian development. It was characterized by an economy-driven approach based on the valorization of cultural capital. It was associated to the liberal ideology (Bianchini, 2001; Landry, 2006a; Lazzeretti, 2012; Porrello, 2006; Stevenson, 2014).

Under some aspects, cultural planning represents an expansive evolution of the planning for the arts (Byrne, 2012; Grandi, 2010; Porello, 2006; Stevenson, 2014).

After the World War II and until the 1970s, the cultural landscape was characterized by the re-launch of "high" culture through public investments in infrastructures[29] (reconstruction or construction of museums, theatres, concert halls, etc.) and people (theatre companies, orchestras, cultural and educational organizations). The purpose was to complete the welfare-state with an adequate cultural offer and, in the most advanced countries, to widen its demand.

In the 1970s and in the first half of the 1980s cultural policies were influenced by the cultural revolution of 1968 moving the attention to "popular culture" - in antithesis to "high" or "status" culture. Spending for temporary events (exhibitions, concerts, etc.) increased and the cultural life was de-centered to the urban suburbs or in minor centers to interrupt the sovereignty of the cities.

In the second half of the 1980s and during the 1990s, urban policies experienced a crisis. The cuts to public budgets, the rise of new-liberalism and of the *New Public Management* were against the direct intervention of the State in the economy. In the field of culture, public operators tried to involve private subjects in the preservation of cultural heritage (but soon investments shifted toward valorization), in the diversification of cultural offer (business-oriented logic). In both cases, culture was meant as an instrument to re-launch the urban economy, a 'resource' for local development[30].

At the end of the 1990s, cultural policies were influenced by the rise of telecommunication and the affirmation of the knowledge-economy. In the celebration of everyday-life as a crucial factor for the cultural produc-

[29] The localization of the infrastructures has always privileged the city and, above all, the cities and the city centers of great historical and artistic tradition.

[30] Aimed at the amelioration of the image of the city, the qualitative level of the urban environment and, consequentially, of the city's appeal to economic operators of various kinds.

tion, the range of cultural resources has been amplified (Bauman, 2005; Usai, 2010). As a consequence, the definition of culture was much wider compared to the traditional anthropological one. It included a creative component consistent with William's definition (1989, p. 90):

> [Culture is] a general *process* of intellectual, spiritual and aesthetic development; [...] the *works and practices* of intellectual and especially artistic activity [as well as] particular *way of life*.

On these basis sociologists and experts forwarded a correction of the economy-driven approach of the early 1990s according to a wider concept of urban culture, inspired by the principles of diversity and authenticity. Thus, new forms of cultural planning spread. They were linked to the local community but without a disciplinary "compartment" (museums, libraries, theatres, etc.) or without a direct public purpose (preservation of memory, education, learning, etc.).

The creative dimension of this cultural planning was nonetheless source of internal tensions (Stevenson, 2014):

- the contrast with public cultural policies for content, scale and community of reference[31];
- an action range so wide that pushed local governments to question what is art or not expressing aesthetic, social and economic judgments in order to come up with the planning (Goodall, 1995);
- the concept of culture as a process, associated to social, political, and economical purposes, which put together social inclusion and education to citizenship (Landry's civic *ethos*);
- an idea of culture as industry which can, or could, be developed for economic income.

From 2000 until today, cultural policies and cultural planning experienced a further development linked to the change of economic scenario first with respect to the theme of creativity and then with respect to the theme of innovation (Lazzeretti 2012, 2013; Stevenson, 2014). In order to understand this last phase it is necessary to analyze the relation between economy and culture, and to divide it for chronological phases. The first substantial change took place at the end of the 1990s with the shift from preservation to "economic valorization" of artistic and cultural heritage, the development of an economics of culture, and the management of cultural organizations. The main purpose was to place culture among the

[31] Even if public cultural policies are still linked to the artistic and aesthetic dimension ('planning for the arts'), to the community and to the national scale, especially in central and southern Europe, where culture-driven models persist.

factors of production, contributing to the affirmation of traditional cultural fields (publishing, cultural tourism), and *no profit* organizations.

Between 2000 and 2005 the debate progressively shifted on the ability of culture to generate and promote local development: culture became a 'resource' for the construction of culture-driven models of development, like the cultural districts. At the same time, the new concept of creative class displayed the contribution of creative economy and human capital to local development, leading to the adoption of the first economy-driven models in the field of culture. These models were centered on cultural industries and, in urban contexts, they were linked to Charles Landry's idea of creative city. From 2006 the studies on innovation have modified the research perspective in favor of a "cultural valorization of economy". They considers culture as a resource for innovation thanks to its creative absorptive capacity, defined as:

> the ability to transform generic creativity into oriented creativity, able to produce ideas and innovations and to transfer them effectively. It depends on the tacit knowledge built in the creative habitat and by the path-dependence of the creative players, whether they are companies or other economic, non-economic, and institutional players, in clusters, cities or regions. Its effectiveness can only be measured in terms of generated innovations, collective or individual, that are only protected by intellectual property rights (Lazzeretti, 2012, p. 16).

From a spatial point of view, this ability can be encountered within the *new creative milieu*[32] and it can be produced through four main paths:

1. Urban regeneration: city branding, urban renewal, iconic projects, *culture-driven* strategies;
2. Economic renovation of departments, districts, products, chains, and professions;
3. Cross-fertilization within the districts or clusters, among themselves or in time (recovery and re-launch of old districts);
4. Serendipity: casual discovery of uncommon connections among new jobs, diversity of interest, correlation between technologies.

Within these paths, cultural policies and cultural planning appear as an implicit spatial policy, nonetheless able to influence explicit spatial policies as well[33]. This inversion of roles can be found in Pratt (2008) where spatial

---

32 The key elements of the *new creative milieu*, besides the creative ability of culture are: (1) the creative *habitat* as a suitable place to attract the creative class; (2) agglomeration creative economies; (3) creative external elements, meaning the cognitive spillover generated by lateral thinking; (4) lateral proximity of informal environments.

33 For the distinction between explicit and implicit cultural policies, see Ahearne (2009).

creative policies are described in relation to culture and not vice versa. He recognizes three models: the consumerist mode, the process-model and the productive model. The first one is mainly based on the marketing of a specific good or cultural event, which already exists or is expressly created[34]. Richard Florida's theory is part of this category because it focuses on urban milieu to attract the creative class (the *bohemians*) within the cities. It leads to strategies of urban regeneration guided by cultural consumerism, in which culture has an instrumental role.

The second model is presented as a correction of the first one proposing art and culture as instruments and practices for the construction of an inclusive and participative city, as opposed to traditional bureaucracy[35]. The deriving strategies of urban regeneration are connected to cultural planning, since cultural activities find are inserted in a territorial strategy that includes other activities in the environmental, social, and economic fields. Lastly, the productive model is based on creative and cultural industries and on their ability to produce talents and innovative ideas. It is defined by a deep connection with the market and the final user that, in the strategies of urban regeneration, leads to the involvement of art and culture only when they are functional to the realization of a project or one of its phases (Evans, 2005).

To conclude, creative spatial policies, cultural policies, and cultural planning are increasingly close for history and purposes[36] (Byrne, 2012), so much that it is possible to imagine their unification within a unique disciplinary field connected to the studies on innovation[37]. A new fundamental role in this process is played by economic rationality which pervaded all public policies, even though it presents controversial aspects. In the specific case of the relation culture-creativity, for instance, it is worth asking ourselves if creative practices have been introduced in substitution of traditional cultural policies, due to their loss of credibility as symbols

[34] The sponsored goods and events are, in practice, a "unique selling proposition – USP – for the territory or the city of reference. They can exist or be planned and realized with that purpose. In the first set it is possible to count, for instance, traditional events such as the Carnival of Venice or Rio de Janeiro, or the contemporaries architectures of Sydney and Genoa's harbors. In the second group there are, on the contrary, the cultural districts constructed in the framework of programs of regeneration of the 1990s of urban transformation societies in England, France, and Spain.

[35] This model is reconnected to Charles Landry's idea of creative city.

[36] They are both instrumental to the construction of new policies of economic development, they use culture for external purposes (economic and social), they put industrial production and individual talent at the core of production, are concerned by territorial planning and the environment, have a specific urban dimension, are born from strategies of requalification and regeneration.

[37] Especially since the cities came to be the center of the international agenda as development propellers.

and expressions of the national states[38] or due to their hostility to private interventions. In a scenario dominated by a managerial attitude, the creative city seems to exasperate the economic discourse, inheriting only the colonialist dynamics from the previous cultural policies: neglected conditions, private interests, export of cultural models[39] (Byrne, 2012; Pratt, 2008, 2011). The sliding of culture to creativity in policies and in planning could be only a strategy to support and encourage an idea of culture as production (cultural and creative industries). This reminds of the recent past when, in order to support and justify "cultural consumption", there was a shift from the elite definition of 'high culture' to the anthropologic one of 'lifestyle'. In any case, the re-proposition of policies and programs based only on the offer exposes the creative spatial policies to the risk of a 'creative infarct', seemingly to those registered in the field of culture with the consumerist strategies[40].

The adoption of an eco-systemic model which involve the external players (tourists, foreign investors) and the internal ones (residents, local businesses) could help in the evaluation of the demand for creative products.

---

38 Not surprisingly, the disaffection to cultural values is accompanied by disregard for the democratic participation in public policies and the gradual emergence of regional systems in economic planning, policies for innovation, in the territorial competition.

39 Besides post-colonialist logics and economic rationality, the loss of credibility of 'public culture' is also related to a deeply rooted suspicion towards technology and cultural industry, perceived as elitisms, deriving from the idea of 'high' culture, inspired by classical arts (Adorno and Horkheimer, 1947).

40 On the theme see Dieter Haselbach, D. and Klein, A. and Knüsel, P. and Opitz, S. (2012). *Der Kulturinfarkt: Von Allemzuviel und überall das Gleich: EinePolemiküberKulturpolitik, Kulturstaat, Kultursubvention.* München: Knaus

Strategies for Creativity

Policies, Programs, Plans

2.1 Creativity and the historical urban landscape read through the international charters and declarations

International charters and declarations describing the relationship between creativity and historical urban heritage cover a century-long period of time: from the Athens Charter of 1931 to the UNESCO Recommendations for the Historic Urban Landscape of 2011.

These documents are a very important testimony of the progresses accomplished in the disciplinary fields of heritage protection and conservation, archeology, architecture, and urbanism. This is mainly true in the case of the historical urban landscape, which experienced a remarkable evolution in the last twenty years. The 'monument' category, which until the 1970s represented the ideal enclosure for the components of the historical architectonical heritage under protection, has shifted towards a much wider label: "landscape". This was conceived as the result of the territorial transformations produced by nature, human activity and their interaction through time (Council of Europe, 2000; Gambino, 2003; Mautone, 2009; Niglio, 2009; Colavitti, 2009; Settis, 2010; UNESCO World Heritage Centre, 2010). Yet, by widening the 'protection' of these historical urban landscapes, they are put in danger of being harmed in the long run (Carta, 1999). Quoting Gambino "if you want to protect everything, you will not protect anything"[1] (Gambino, 2003). This critique mainly emerges in relation to the need of circumscribing the sectors of the historical city to put under protection according to their landscape value. This is why in the *Operational Guidelines for the Implementation of the World Heritage Convention* of 1992, UNESCO forwarded the idea of the institution of two different types of landscapes, then gone over in the guidelines of 2008. The first one is reserved for cultural landscapes, fruit of the interaction between man and nature, as it is described in the European Landscape Convention – ELC (2000). The other one concerns historical towns and town centers.

1 In the original text: "se si vuol proteggere tutto, non si proteggerà più niente" (Gambino, 2003).

Nonetheless, this choice puts UNESCO in an antithetic position in respect to the ELC which has established, once and for all, that landscape, as a good of the community, is to be protected, managed and/or designed independently from its concrete value (Priore, 2005). Between 2002 and 2005, UNESCO proposed to update the principle of the Charter of Venice and the diffusion of a new and extensive approach in order to reduce the conflict. This was to provide an effective application of the ELC, keeping in mind the increase of global urban population (OECD, 2006). The *World Heritage Cities Program*, launched in 1996, was revised and updated. In 2002 in Urbino the *Partnerships for World Heritage Cities. Culture as a Vector for Sustainable Urban Development* convention took place. During this event, the first mapping of historical towns and monuments in the urban field was carried out along with the first recommendations for historical towns (UNESCO World Heritage Centre, 2003b). In order to make the confrontation on cities and cultural landscapes easier, the Memorandum of Vienna was written. The Declaration for the Preservation of Historical Urban Landscape was drafted based on this document, and was adopted in October of the same year and in addition to this, the *Historic Urban Landscape (HUL)* project was launched. It was characterized by the intense involvement of the territory (UNESCO World Heritage Centre, 2010), thanks to which a first agreement on the definition of *Historic Urban Landscape* (2008) and on the adoption of the *HUL Recommendation* (2011) was reached. According to the recommendation the architectural heritage of the city was brought together and the "historical urban landscape" was defined as:

> The urban area understood as the result of a historic layering of cultural and natural values and attributes, extending beyond the notion of "historic center" or "ensemble" to include the broader urban context and its geographical setting.
>
> This wider context includes notably the site's topography, geomorphology, hydrology and natural features; its built environment, both historic and contemporary; its infrastructures above and below ground; its open spaces and gardens, its land use patterns and spatial organization; perceptions and visual relationships; as well as all other elements of the urban structure.
>
> It also includes social and cultural practices and values, economic processes and the intangible dimensions of heritage as related to diversity and identity (UNESCO, 2011).

This is a broader definition that finally identifies the already existing city as a landscape recognizing its peculiarities. It synthetically expresses the cultural evolution that in the last two centuries involved the attribution of the value to the signs of the past, and that lead to the recognition of monuments, then to historical buildings and, at last, to the urban landscape as a whole.

The social and economic role of the historical urban landscape as a premise for creativity was investigated in the 1960s along with the economic theories on the knowledge-based society[2] (Andersson *et al.*, 2011; Usai, 2010). In this sense, the European Charter of Architectural Heritage and the relative Declaration of 1975 represent the first international document of reference. The existing city is recognized as a fundamental part of the life of the people and, therefore, as a *common good* of public interest. By defining the city as a common good, every citizen is recognized the right to have an active role in the decision-making process concerning the preservation of the multilayered historical town. Consequently, the Charter and the Declaration of Amsterdam supported a new and dynamic approach to preservation, where learning, participation, and strategy go together allowing local communities and private players to put their "civic creativity" into practice following a list of rational and shared rules (Landry, 2006a).

The "*Recommendation on participation by the people at large in cultural life and their contribution to it*", signed in Nairobi one year later deepened the concrete outcomes in the exercise of "civic creativity" on behalf of the local communities and private subjects confirming that (art. 3):

a) the concept of culture has been expanded to include all forms of creativity and collective or individual expression (lifestyle and artistic production);
b) a free and democratic access to culture for all the people presupposes the existence of suitable economic and social policies;
c) the participation to cultural life presupposes the involvement of different social partners in the definition of sector policies, and in the managing and evaluation of the interventions as well.

According to the Recommendation, this can be achieved through administrational decentralization, territorial animation, communication, support for artistic production and concentration/cooperation on a local level[3]. The Convention for the protection of the European architectural heritage (Granada, 1985) takes over these concepts, yet it reverses the point of view on

2 Given the existing connection between creativity and culture, the timing shows several similarities with the one redacted for cultural policies and planning.

3 The idea of cultural heritage as expression of human creativity can be found in the *Operational guidelines for the implementation of the world heritage convention* of 1977. The first criterion for the recognition of the universal value of a property and its enlisting in the UNESCO World Heritage List, in fact, is "representing a masterpiece of human creative genius".

culture and creativity. In Nairobi's Recommendation, Culture is described as expression of creativity whereas in Granada's Convention it is defined, through the architectural heritage, as an "element of identity, inspiration and creativity for present and future generations" (art. 15). The Convention concentrates on education policies focusing on cultural heritage and its link with the architectures, arts, lifestyles and folklore of Europe.

In following years the entrance of emerging countries in the UNESCO World Heritage List moved the debate on the methodologies adopted to identify, preserve and manage the UNESCO's sites[4]. In this perspective, the 1994 Nara Charter supported cultural diversity as a valuable part of the UNESCO World Heritage and included the *verification of authenticity* as a mechanism for the recognition of UNESCO's properties in order to avoid subjective opinions in assigning value. "Authenticity" is intended as the *integrity* of the property in various aspects[5] (see Table 4). However, the distinctive elements in the recognition of the property still remain those related to the creative process which include: the ideation by the artists, the creation of the artifact and its history. This is according to Cesare Brandi's theory of restoration (Jokilehto, 1995, 2008).

Table 4. Criteria for evaluation of the integrity of the asset

Location	The place where the historic property was constructed or the place where the historic event occurred (important to understanding why the property was created or why something happened).
Design	The combination of elements that create the form, plan, space, structure, and style of a property.
Setting	The physical environment of a historic property, i.e. the character of the place in which the property played its historical role. It involves how, not just where, the property is situated and its relationship to surrounding features and open space.
Materials	The physical elements that were combined or deposited during a particular period of time and in a particular pattern or configuration to form a historic property.

4 The previous methodology was based on the aesthetic and formal canons of the Western world. It was questioned and revised to face the cultural models of the Muslim and Buddhist countries and those of the indigenous populations in postcolonial states. On this topic, see *Global Strategy for a Representative, Balanced and Credible World Heritage List*, launched by UNESCO in 1994, and the ICOMOS evaluation *Filling the Gaps: an Action Plan for the Future. An Analysis* written in 2004.

5 They come from the American procedure for the registrations of buildings and spaces to the National Register of public spaces (National Park Service, 1995).

Workmanship	The physical evidence of the crafts of a particular culture or people during any given period in history or prehistory. It is the evidence of artisans' labor and skill in constructing or altering a building, structure, object, or site.
Feeling	A property's expression of the aesthetic or historic sense of a particular period of time.
Association	The direct link between an important historic event or person and a historic property.

Source: National Park Service (1995)

The 1994 Charter of Aalborg applied the principles of authenticity to the urban landscape[6] expressing the uniqueness of the historical cities and the need for context-specific sustainable development models, which derive from a collective creative process (point I.4).

Within two years of the Nara Charter the progressive connection with the theories of local development that see architectural heritage as a tool to launch urban economy[7] lead to the publication of the first UNESCO report: *Our Creative Diversity*.

In this document, creativity is considered an instrument to encourage responsibility among authorities and local communities, in order to compete with large cultural industries (publishing, cinema, etc.). They are called to systematically map the social and technological resources present in the local field (territorial capital), to elaborate alternative methods of development and new public policies, to obtain the transfer of functions from national governments, and to create networks of relations between the territory and the global market. Within cultural policies, this means to redefine the preservation field (still focused on "high culture"), to update the methodologies for the recognition and the protection of the assets, to boost museums and other cultural institutions, to include and raise the awareness of local communities towards their own heritage. These policies are aimed at a sustainable development keeping in mind the risks of "mass" tourism.

Lastly, the report invites local organizations to protect, support and sponsor traditional and innovative productions (pluralism of information, protection of copyright and intellectual property) (World Commission on Culture and Development, 1995). The 1999 Charter of Burra, the 2000 Charter of Cracow and the 2000 European Landscape Convention (ELC) proposed the insertion of culture among the factors of sustainable development. They also promoted landscape as a fundamental element for

6 As expression of long lasting, specific and non-reproducible processes.

7 A resource used in cultural planning in the late 1990s to correct the economy-driven approach of the previous period.

well-being and for the quality of life of citizens and as an attractive factor for the creative class and for local talents. According to this, Landry's creative *ethos* can be redefine as "landscape creativity", indicating the set of acts, ideas or products, which allow modifying an existing landscape or generating a new one through the re-launch of civic ethic. This approach is based on the idea that each generation has the right to "leave a trace of its culture on the landscape and make it suitable to its needs" (Ronza, 2009, p. 206). Anyways, it was a sign of a diffuse creativity, different from a creativity oriented to innovation, which have characterized the next phase.

In the early 2000, globalization stimulated cultural international organizations to update their principles and the prerequisites for their work still dependent on a Western vision. The UNESCO re-opened the debate on cultural diversity introduced by the 1994 Document of Nara, adopting three fundamental documents on historical urban landscape and creativity:

1. the Universal Declaration on cultural Diversity of 2001[8];
2. the Agreement for the protection of non-material cultural heritage of 2003[9];
3. the Agreement on protection and promotion of diversity of cultural expressions of 2005.

Earlier analyzed as a component of the individual creative process (source of inspiration for artists and in planning for the arts), or of the collective one (Landry and Florida's creative *ethos*). In these documents creativity is connected to cultural production with an explicit reference to cultural and creative industries and to the themes of social cohesion and international cooperation. *Cultural expressions*[10], result and source of creativity, are considered an essential element for the collective well being and urban livability. Furthermore, they are a necessary component of public policies because without them the production of new tangible and intangible heritage for future generations would not be possible.

The connection with production is symptom of the detachment from Bianchini and Landry's models, based on "local resources", and of a progressive alignment between cultural and creative policies. Above all, the analyzed

---

8 See art. 1, art. 7-12.

9 See art. 2 comma 1, art. 14, 15, 19.

10 Cultural expressions mean expressions of cultural content, deriving from the creativity of individuals, groups of people and societies. The activities, the assets, and the services that represent or transmit cultural expressions for their qualities, their usage, and their specific purpose, are defined as cultural independently from the commercial value attributed to them. The industries producing and distributing cultural assets or services are defined as cultural as well. See art.4 comma 3, 4 and 5 of the Convention on the protection and promotion of diversity of cultural expressions (2005).

international declarations evidenced the relation between culture, creativity, and innovation, stating a reference to creative clusters and networks and on cultural and creative industries. In particular, they emphasize the role of generic and oriented invention within the cultural heritage chain, made of: safeguard, knowledge and diagnosis, protection and preservation, valorization and fruition, management (Amari, 2006; Finnel and Santagata, 2007; Mautone, 2009) (see Figure 2). This emerges overall in the charters on architectural heritage, especially in relation to the sustainability and the landscape creativity. Despite this, they do not introduce any significant novelty respect to the previous charters. Therefore, on the basis of Vienna's *Memorandum* of 2005, UNESCO opened a series of consultation concerning the instruments for the preservation and management of the historical urban heritage. The outcomes of the consultations were integrated in the 2011 Recommendation on the Historical Urban Landscape. The recommendation represents a relevant step forward in the study of landscape and cultural heritage as it applies the ELC approach to cultural planning (points 2, 5 and 15). In addition to this, the HUL Recommendation recognizes cities as propellers of development thanks to their role of hub or relational network in innovation processes (point 1). It refers to the most recent theories on the creative capacity of culture and on the techniques used in the creative spatial policies. For instance, cultural mapping, the use of participative techniques, the inclusion of cultural planning into a global and integrated urban strategy[11], the construction of adequate partnerships to reach the project goals[12] (van Oers, 2010).

Figure 2. *Cultural heritage chain*

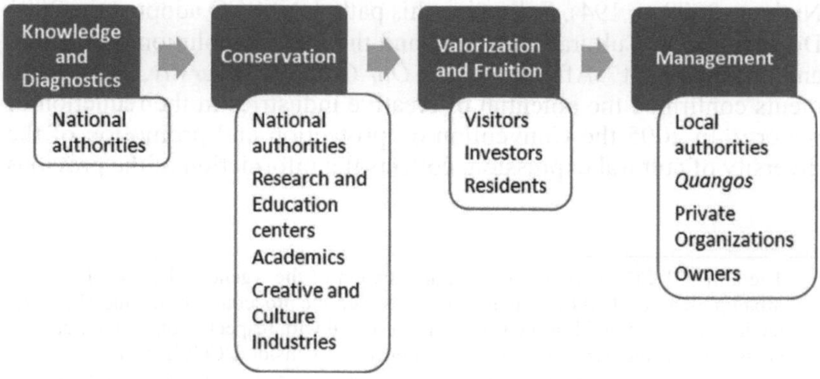

Source: elaboration of the author

[11] This has been long desired in urban planning as well, which, in the last years, is trying to fulfill this purpose through the strategic planning of European funds.

[12] See § IV point 5; art. 22, 24.

2.2 UNESCO programming for creativity and culture

The 1990s represent a breakthrough with respect to the previous political and programmatic phase. The opening to the global market and emerging Countries forces a revision of the system of values at the base of international policies for development and culture, still referred to the Western view. The main issue was to substitute the capitalistic model by generating sustainable forms of production and consumption for resident communities rise. A first answer arrived in 1992 with the Declaration of Rio and the Agenda 21, i.e. a plan of actions by the United Nations, which shifted the concept of sustainable development to a local scale through interventions for economic growth, social inclusion and environmental assessment (the three 'pillars'). Despite the two initiatives did not have the desired effects, they led local administrators to collaborate within formal and informal networks. This is the case of the *United Cities and Local Governments* (*UCLG*) that in 2004, following different declarations of UNESCO, elaborates an Agenda 21 for Culture, structured on the following topics: human rights, governance, sustainability and territory, economy and social inclusion. Culture was faced in all its shades, from creativity to lifestyles, as a transversal asset in all the themes of the UCLG agenda[13] (Stevenson, 2014). After years of research, even the UNESCO recognized the increasing interest towards new sustainable and successful development models in which cultural policies can be framed. During the Intergovernmental Conference on Cultural Policies for Development, held in Stockholm in 1998, UNESCO claimed for an international policy framework on culture and cooperation[14] (United Nations, 2008, p. 194). Following this path, UNESCO adopted the 2001 Declaration on Cultural Diversity, and the 2003 Resolution on Culture and Development[15]. After the report *Our Creative Diversity*, these documents confirmed the potential of creative industries in the reduction of poverty. In 2005 the Convention on protection and promotion of the diversity of cultural expression, collects the information of the previous

---

13 The aim of UCGL network was the application of the Agenda 21 principles to the urban policies drafted by its members. However, the procedure for the members' selection is unclear and Europe is more represented with respect to other geographical areas, such as the Asian one, in terms of candidates inside UCGL's board.

14 It is worth remarking the program for *Arab Capitals of Culture*, launched by UNESCO in 1995 as a part of the *Arab Plan* and implemented with the *Arab League's Educational, Cultural and Scientific Organization* (ALECSO). Along time, it assumed a political connotation facing times such as fight, migration, resistance in the Arab and Islamic culture (almost as opposed to the program for the European Capitals of Culture) (Stevenson, 2014).

15 Resolution of the General Assembly n.57/249 of 20th February 2003.

documents reaching the formulation of three programmatic purposes characterizing the UNESCO activity[16]:

1. free access to culture through pluralism of information and media;
2. mobility of the creatives through institutional programs and partnerships for social cohesion and international cooperation;
3. fair remuneration of creative work through: protection of copyright and the rights connected to it, access to credit for artisans and small and medium businesses (with respect to mass productions and cultural and creative industries).

Starting from these goals, UNESCO tried to integrate cultural industries in the international agenda and promote the artisanal and informal component in the field, activating different collaborations, alliances, and networks. In 2004 an informal multi-agency group on creative industries was established at the United Nations. It gathers twice a years in Geneva, and it is constituted by representatives of UNESCO, *International Labour Organization* (ILO), *International Trade Centre* (ITC), *United Nations Conference on Trade and Development* (UNCTAD), *United Nations Development Programme* (UNDP)[17] and *World Intellectual Property Organization* (WIPO) (United Nations, 2008).

The UNESCO also launched the *Creative Cities Network*, a "network of networks" operating in different fields: literature, cinema, music, design, gastronomy, media arts, handicraft, and folk art[18]. Compared to Florida and Landry's standard solutions, the UNESCO network presents a new concept of creative city, characterized by a more nuanced approach, sensitive to cultures and local differences, far from the classical strategies of urban and territorial marketing. The purpose is the construction of local alliances, sharing experiences and challenges that characterize the contemporary city[19] (Stevenson, 2014; United Nations, 2010, p. 15). Important

16 See the "*Guidelines for an Action Plan*" of the Declaration on Cultural Diversity (2001), paragraphs 14-19 and the Declaration on Cultural Expressions (2005), art. 6 paragraph 2.

17 Since 2004 UNDP hosts a *High-level Panel on Creative Industries and Development.* In 2006 the same organization has launched the *Creative Economy and Industries Programme.* The well-known UN's reports on creative economy are published thanks to this program.

18 It is not easy to find information on the subject and today the main source of reference remains the webpage of UNESCO: http://www.unesco.org/new/en/culture/themes/creativity/creative-cities-network/

19 According to Pratt (2011, p. 22) the main challenge for the future of the networks consists in the constitution of analytic models that treat cultural consumption and production together, on the base of recent studies which proved the specific and contextual nature of creativity.

are the references to collective learning and to cities as creative hubs or socio-cultural clusters that witness the inclusion, on UNESCO's side, of the theories on innovation and on creative absorptive capacity.

Among other initiatives supported by UNESCO there are the symposium that took place in Nagaur in 2005[20] and the second meeting of the Ministers of Culture in Africa, Caribbean, and Pacific area (ACP), held in Santo Domingo in 2006. In this occasion, the project *Strengthening the Creative Industries in Five Selected ACP Countries*, later approved in 2007, was proposed. It was the first project implemented jointly by different international organizations (ILO, UNCTAD, and UNESCO) with funds coming from the European Commission and the ACP Secretary[21].

In 2008 the United Nations set up several long lasting initiatives in collaboration with other organizations on creative economy (statistic models) and on cultural industry (cooperation with emerging Countries, intellectual property). The first was the initiative *Creative Africa* and the *Creative Economy Report* thanks to the joint work of UNCTAD and the Special Unity for South-South Cooperation of the UNDP[22]. It developed an international statistic framework for cultural industry thanks to the data from different agencies[23]. The same year a direct collaboration between UNDP and UNESCO was set up. It was dedicated to the activation of eighteen national programs funded by the "Thematic window on Culture and Development" of the *Millennium Development Goals Achievement Fund*[24]. The purpose of these initiatives was to bring to the surface the expertise that

20 The recommendations of this event, best known as *Jodhpur Initiatives*, proposed a series of strategies for the gathering of national data and the development of the creative industry in Asian Countries (United Nations, 2008, p. 15).

21 The countries involved in the pilot project were Fiji, Mozambique, Senegal, Trinidad, Tobago, Zambia (United Nations, 2008, p. 46).

22 The official launch of the initiatives took place within the 12th session of the four-year ministerial Conference of the UNCTAD in Accra, Ghana, from 20th to 25th April 2008 (United Nations, 2008).

23 In 2006 the UNESCO Institute for Statistics (UIS) had started a survey concerning the creation of a unitary statistic framework on cultural and creative professions. See Burns Owens Partnership, Pratt, A.C. and Taylor, C. (2006). *A framework for the cultural sector: A report for UIS/UNESCO*, London, BOP.

24 The *UNDP-Spain Millennium Development Goals Achievement Fund (MDG-F)* represents an innovative approach to international cooperation with the purpose to accelerate the fulfillment of the Millennium Development Goals and to sustain the reform of the United Nations. With its launch in December 2006, the UNDP and the Spanish Government created an important partnership with a total investment of 918 million euros destined to the funding of national governments, local authorities, and the organizations of civil society in the battle against poverty and inequality. The MDG-F supports eight "thematic windows" among which one is dedicated to Culture and Development. UNESCO, as specialized agency of the United Nations with a specific mandate on culture, was awarded coordinator of this thematic window (United Nations, 2013).

guides the creative and cultural initiatives worldwide. Always in the MDG-F framework, UNESCO launched the *Framework for Cultural Statistics 2009* (FCS) and elaborated an operational methodology based on cultural indicators for local development. Thanks to the support of the Spanish government the methodology was applied in eleven countries[25]. In 2009, the Departments for Cultural Policies and the Intercultural Dialogue of UNESCO signed an agreement with UCLG's Culture Committee to produce a survey explaining the point of view of cities and local administrators in the process "*Towards a new model of cultural policy*". The Report n.4, *Culture and Sustainable Development: Examples of Institutional Innovation and Proposal of a New Cultural Policy Profile* by the Agenda 21 for Culture, represents the outcome of the agreement[26].

In the period 2009-2011, the Supporting Program for Creative Industries, implemented by UNESCO, ILO, FAO and UNDP along with the Kingdom of Cambodia, trained more than 800 artisans and artists. Later, from 2010 to 2013, the International Fund of UNESCO for Cultural Diversity organized four cycles of planning destined to emerging countries for a total of 71 cultural financed projects[27] (United Nations, 2013, p. 134-135). At the same time, the Program for Statistics on Culture of the UNESCO Institute of Statistics (UIS) supported the diffusion of sector statistics on a global level through uniformed and comparable data, shared methodologies and standards, training courses and technical support to national statistic institutes. These activities prove an increasing involvement of UNESCO in the creation of a partnership building strategy, which was mainly addressed to the other agencies of the United Nations, but was also open towards the contribution of other international organizations such as ALECSO or UCLG. The strategy was aimed at encourage the closeness of international organisms and national and local governments on the themes of sustainable development and culture in view of a major reform of the UN's bodies.

---

25 The initiative of *Culture for Development Indicators Methodology* (UNESCO CDIS) wanted to contribute to the actuation of article 13 (Integration of culture of sustainable development) of the Convention on cultural expressions (2005). The national projects started in 2009 and lasted three or four years. They concerned the following countries: Bosnia Herzegovina, Burkina Faso, Cambodia, Colombia, Ecuador, Ghana, Namibia, Peru, Swaziland, Uruguay and Vietnam (United Nations, 2013, pp. 167-168).

26 See Cultural Policies and Sustainable Development, www.agenda21culture.net [on line]. Available on http://www.agenda21culture.net/index.php/it/16-official-documentation-all/agenda-21-culture-all/437-cultural-policies-and-sustainable-development [last access: 19th March 2014].

27 See International Fund for Cultural Diversity (IFCD), www.unesco.org [on line]. Available on: http://www.unesco.org/new/en/culture/themes/cultural-diversity/diversity-of-cultural-expressions/funded-projects// [last access 19thMarch 2014].

The UN Resolution on "Culture and Development"[28] (2011) and the UNESCO Declaration of Hangzhou "*Put culture at the core of sustainable development policies*"[29] (2013) can be considered the first results of the UN's reform.

The three goals introduced by the *Convention of protection and promotion of the diversity of cultural expressions*, taken over by the European Union as well, constituted the fundament of the UNESCO activity the following ten years. They characterizes the modalities with whom the organization decline the concept of "creative capacity of culture" through their programs and projects. An increasing importance is given to partnership building strategies, conceived as the set of links that organizations establishes or wants to establish with local and *supra*-local players through formal agreements or joint ventures. Their purpose is to guarantee substance and, therefore, a symbolic value to the competitive identity that policy makers and planning authorities want to build and develop for a specific place (Hankinson, 2010). The UNESCO strategy covers initiatives oriented on four main themes: culture, sustainability, and local development (with particular attention to international cooperation). In spite of this, spatial creative policies are hardly discussed, perhaps limited to the *Creative Cities Network*, to the projects for statistic and spatial mapping of creative and cultural industries, to the mapping of local resources in the creation of creative districts. On the other hand, importance is also attributed to the political action of UNESCO, in terms of coordination and alignment of cultural national agenda with respect to the global aim of sustainable development[30] (see Figure 3 and 4).

---

[28] UN Resolution n.66/208.

[29] The Hangzhou Declaration contains ten key points for a sustainable development (United Nations, 2013, pp. 156-159): (1) Recognizing that creative economy does also produce a monetary value, which significantly contributes to reach an inclusive and sustainable development, in addition to its economic benefits; (2) Making culture a trampoline for the processes of economic, social, and environmental development; (3) Revealing the existing opportunities through a mapping of creative economy's local resources; (4) Strengthening the base of knowledge through a rigorous collection of data before every policy on creative economy; (5) Investigating the connections between informal and formal sectors of creative economy; (6) Analyzing the critical factors of success which contribute to the creation of new paths for the local development of creative economy; (7) Investing in the sustainable development of creative companies all along the productive chain; (8) Investing in the strengthening of local abilities; (9) Engaging in the South-South cooperation to ease reciprocal learning and define international agendas for development; (10) Making culture prevail in the programs of economic and social development, even with respect to concurrent priorities.

[30] As it can be expected being UNESCO an operative branch of the United Nations.

Figure 3. *The fields of action of UNESCO: organizations and activities from 2004 to 2014.*

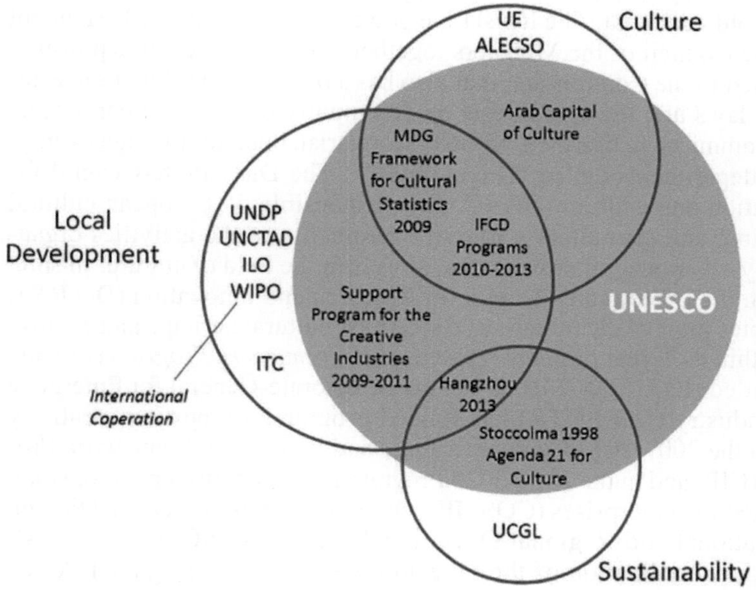

Source: elaboration of the author

Figure 4. *MDG-F Portfolio: distribution of resources for activity.*

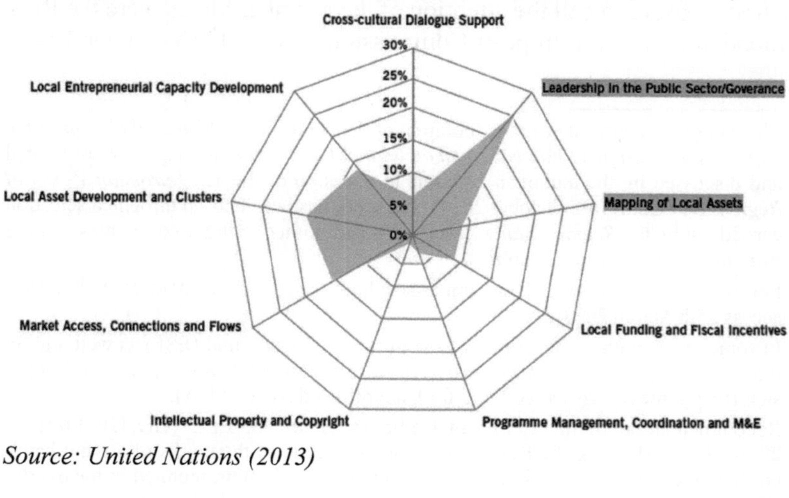

Source: United Nations (2013)

2.3 European programming for creativity and culture[31]

The European Union (EU) is a legal Union of twenty-eight countries of the old continent. The legislative power is assigned to the Parliament and the Council of the Ministers together, while the executive power is assigned to the Commission that also has a propositional function as far as the laws and the agreements on community policies are concerned. The Commission, like the UNESCO secretariat, operates through twenty-eight departments called "Directorates"[32]. The Directorate-General for Education and Culture (DG EAC) is responsible of European cultural planning, and guarantees a role of coordinator on the activities organized by other general directorates, always in the field of culture. In spite of that, the Directorate-General for Research and Innovation (DG R&I) promotes most of the initiatives concerning cultural heritage and creativity within its 7-year program – first named *Framework Program* (FP) and now titled HORIZON 2020. Even the Directorate-General for Enterprise and Industry (DG ENTR) has funded programs supporting creativity within the 2007-2013 Competitiveness and Innovation Framework Program (CIP) and in the 2014-2020 Program for Competitiveness for small and medium enterprises (COSME). Other initiatives are realized though the National and Regional Operational Programs (NOPs and ROPs), with the coordination of the Directorate-General for Regional Policy (DG REGIO) and the resources of Structural Funds, in particular those of the European Regional Development Fund (ERDF)[33] (Chapuis *et al.*, 2009; Kern and Montalto, 2013).

Seemingly to what happened on a global level, the European cultural planning is rooted in the protection and preservation of tangible heritage. The first measure to call the attention of the Member States were the Recommendation of the European Commission dated 20th December 1974[34]

---

31 The paragraph constitutes a re-elaboration of the contribution "*Cultural planning: the impact of the European Funds System on regional and urban development*" presented and discussed by the author in Brussels in occasion of the 12th *European Weeks of Regions and Cities* (6-8 October 2014). The work has been chosen through a selection carried out by the *Region Studies Association* and limited to thirty contributes coming from different European Countries.

32 For the complete list of the Directorates see: http://ec.europa.eu/about/ds_en.htm [last access 25th March 2013].

33 In some cases, with the resources of the European Social Fund (ESF) as well and, in those countries of the Convergence objective, of the Fund for Development and Cohesion (FDC), previously called Fund for Underutilized Areas (FUA).

34 Recommendation of the European Commission n.75/65/EEC (GU UE L021 of 28/01/1975). Also, see the Resolution on conservation of the works of art and handcrafted products of the Ministers responsible of cultural affair, reunited in the quarter of the Council of 13 November 1986, and the Resolution of the Ministers responsible

and the Resolution of the European Parliament to the Member States on urgent measures for the preservation of architectural and natural European heritage, considered a key factor in the quality of life of European citizens. Keeping in mind the European Treaty and other international agreements signed by EU, the two aforementioned measures established the subsidiarity of EU decisions and programs with respect to the laws adopted by the member States themselves.

This is one of the reasons why, on a European level, it is not possible to talk about public policies by the EU in the field of culture. Instead, it is possible to talk about cultural planning, i.e. a coordinated implementation of strategic and coordinated initiatives aimed at the constitution of a European identity.

In 1983, the European Commission started a first series of operative and concrete measures connected to the cultural heritage, providing its sustain to the restoration of prestigious sites and monuments, such as the Acropolis of Athens and the monastery of Mount Athos in Greece. Between 1984 and 1995, the Commission supports a series of pilot projects for the preservation of architectural heritage within the Framework program for Research and Innovation of the DG R&I[35] (Chapuis *et al.*, 2009). At the same time, the initiative "City of European Culture", then renamed "European Capitals of Culture" (ECoC), was established in order to promote the richness and diversity of European cultures, to celebrate the cultural bonds connecting Europe, to lead the citizen to the knowledge of member States, to spread a feeling of European belonging. Thanks to the ECoC program in 1986 the network EUROCITIES was established. In particular, it deals with: the creation of supportive actions for cities; urban life; experiences and priorities in city administration (Stevenson, 2014).

Before 1992, culture was not included in any of the three foundational pillars of the European Community. It was thanks to art.128 of the Treaty of Maastricht (1993) and, later on, to the art.151 of the Treaty of Amsterdam (1997) that cultural heritage found a place in the first pillar of EU action, dedicated to community policies (Cirillo, 2003).

In 2000, the Strategy of Lisbon proposed to make Europe more competitive within 2010 fostering the knowledge economy and the human capital. The strategy proposed reforms for the economic development and employment, for the research, education and permanent training. It supported culture as a catalyst of creativity and innovation, above all in regional policies (Chapuis *et al.*, 2009, p. 5; MiBAC, 2007, pp. 35-37;

of the cultural affairs and the protection of the European architectural heritage (GU UE C320 del 13/12/1986).

35 In the program of environmental research (*Environmental Research Theme*).

Melina, 2011). In 2001 Tony Bennet studied the European programs on culture recognizing the positive influence of EU on at least three factors that are crucial for cultural diversity: internal mobility, cultural integration, social and cultural change (Bennet, 2001). Cultural and creative industries (CCIs) were included in European cultural planning to support cultural diversity[36]. Consistent with the European Parliament, several institutions and organizations push for a common policy on CCIs within a broader European strategy for culture (MiBAC, 2007, p. 35-37). Between March 2003 and March 2005, for instance, the EUROCULT21 network took on the construction and the diffusion of adequate professional and administrative competences on a local level. The initiative led to a shared framework for the draft of urban cultural policies[37] (Stevenson, 2014, p. 127). The first conference organized by the European Commission, entitled "*Sustaining Europe's Cultural Heritage: from research to policy*" took place in London in October 2004 (Chapuis *et al.*, 2009, p. 21; JPI Cultural Heritage, 2010). In 2006 the first study on culture economy was published (KEA, 2006). The mechanism of selection of the European Capitals of Culture[38] was institutionalized and the Commission held a conference in Prague, waiting for the initiative *Safeguarded cultural heritage: understanding and viability for the enlarged Europe* (SAUVEUR). In view of the 7th Framework Program (FP7) of the DG R&I, the conference focused on the impact of EU projects (results and feedback received by the stakeholders). In particular, foreseeing a reduction of funds for cultural heritage in the FP7 (in favor of creativity and innovation), the final document of the conference invited the EU authorities to (Chapuis *et al.*, 2009, p. 21):

- include the themes of cultural heritage among the priorities and the purposes of FP7 and intensify the support to research on cultural heritage within the same field;
- create tools to overcome fragmentation in research and ease the access to funds by the experts in the cultural heritage field;
- guarantee that in the revision of the Strategy of Lisbon of 2009 the value of research on cultural heritage is explicitly mentioned;

---

36 Resolution of the European Parliament on cultural industries (P5_TA(2003)0382).

37 The so called "4E model": *enlightenment, empowerment, economic impact, entertainment.*

38 Decision n.1622/2006/CE of the European Parliament and decision of 24th October 2006 of the Council institute the event "European Capital of Culture" for the period 2007-2019 with a mechanism of rotation for the member countries and a procedure of selection provided by the European institutions (earlier on, the selection was provided by the Council of the Ministers of Culture).

According to the goals of the 2005 UNESCO Convention on cultural expressions 2005 and the recommendation of the Conference of Prague, the European Commission included culture among the *European Research Areas*[39] of the FP7 by the DG R&I (FP7). This implied the creation of a specific agenda for culture, as it is reminded in the Communication of the Commission to the European Parliament dated 10th May 2007, concerning a "European agenda for culture in an increasingly globalized world"[40] (Chapuis *et al.*, 2009, pp. 8-9; MiBAC, 2007, pp. 37-39). In November 2007 the Council of the Ministers of Culture approves the European agenda for culture[41]. The Council's resolution includes the three fundamental objectives of the Communication of May 2007 and five priorities for the period 2008-2010. One of these concerned the access to culture through the promotion of cultural heritage, cultural tourism, use of multiple languages, digitalization, synergy with education (artistic education) and a major mobility of the collections (JPI Cultural Heritage, 2010). The European Agenda for culture was put into practice through:

1. the draft of a *vision document*;
2. the creation of a strategic agenda for research activities;
3. the implementation of this agenda through joint programming initiatives – JPIs.

The Conclusions on the joint programming for research in Europe in 2009[42] included "Cultural heritage, climate change and security" in the topics of the FP7 (period 2009-2010)[43].

Others initiatives were financed by the DG EAC, DG ENTR, DG REGIO. For instance, the *European Creative Industries Alliance* (ECIA)[44] created in 2012 by the DG ENTR with 6,75 million euros of the *Competitiveness and Innovation Programme* (CIP)(Kern and Montalto, 2013;

---

39 European Commission (2007). European Research Area: New Perspectives – The "Green Paper", COM(2007)161, Brussels, 4 April 2007

40 Communication n. COM(2007)242def (Not published by the official news).

41 Resolution n.207/C 287/01 del 16 November 2007

42 Conclusions n.17226/09 RECH 454 SAN 359 adopted by the Council for Competitiveness in occasion of the meeting held in Brussels on the 3rd December 2009.

43 The *vision document* of the initiative of joint programming on cultural heritage and global change (*Joint Programming Initiative on Cultural Heritage and Global Change* - JHEP) was adopted on the 17th June 2010.

44 ECIA, The European Creative Industries Alliance: What? Why? How? Who?, www.eciaplatform.eu [on line] Available on: http://www.eciaplatform.eu/newsarticle/ecia-in-a-nutshell/ [Last access: 28 march 2014]

Palmi, 2013). Again, the *European Creative Districts* by the DG ENTR and the European Parliament (the first regions awarded were Vallonia and Tuscany).

The DG REGIO was directly responsible of the LEADER and URBACT programs and it coordinated the projects implemented by national and regional governments via NOPs and ROPs. Among the projects founded by the DG REGIO and realized with ERDF funds, there are the project *Creative Cluster in low density urban areas*, part of the URBANACT program, and the *Creative Cities* project, part of the *Central Europe Programme* (Kern and Montaldo, 2013; INTELI, 2011). Lastly, the event "2009 Year of creativity and innovation" launched by the European Commission dealing with several projects[45] (MiBAC, 2007, p. 37-39).

On the 17th of June 2010 the European Council approved the strategy Europe 2020 which substituted the contested Lisbon Agenda. The document establishes the principles and the priorities for the spending of EU funds in the 2014-2020 programming. The strategy intends to accelerate the recovery from the present economic and occupational crisis. Five are the strategic objectives of Europe 2020:

- 75 % of people between 20 and 64 years will have a job;
- 3 % of the PIL of the EU will be invested in R&D;
- The goal "20/20/20", about climate/energy will be accomplished (included an increment of 30 % of the reduction of emissions, if the conditions allow it);
- The rate of school drop-out will be lower than 10 % and at least 40 % of the young will have a graduate degree;
- 20 million people less will be at risk of poverty.

This is why the document points out three priority axes for the Cohesion Policy 2014-2020: *intelligent growth* (research, competence and innovation); *sustainable growth* (efficient use of resources, reduction of carbon emissions, integrated industrial policy); inclusive growth (occupation and education) (Table 5). Making their own the recent theories on innovation, the European Union adopts an "indirect" approach, based on the creative capacity of culture which, in a systemic approach, finances the programs of research and development to re-launch culture instead of initiatives directly dedicated to this one (KEA, 2012; Kern and Montalto, 2013; Palmi, 2013). The NOPs and the ROPs, funded by ERDF and ESF, are still yet to come but the regulation for their draft are already

[45] See the *2009 Create* project at: http://www.create2009.europa.eu/projects.html [last access: 28 March 2014]

available. According to the DG REGIO's provision on the 2014-2020 ERDF, the Policy of Cohesion will host some programs dedicated to the cultural heritage, as an answer to the priority n. 6 of Europe 2020, but they will represent a little quote if compared to the indirect funding to culture, deriving from priority n. 1, 8, and 10 of the strategy. The latest projects of the 2007-2013 programming and the programs of the other Directorates complete the frame of the future European cultural programming (see Figure 5).

Horizon 2020 is the new program of the DG R&I, alongside with the latest initiatives of the FP7 for 2007-2013. The funding are provided according to three fundamental parameters: scientific excellence (research); industrial competitiveness (ITC, nanotechnologies, robotics, biotechnologies and spatial research); Europe in a changing world: challenges for an inclusive, innovative, and reflexive society.

The calls for proposal dedicated to culture satisfy mainly the third requirement, concerning the constitution of a more inclusive society, thanks to socio-economic and humanistic research on the themes of identity and European history (as in FP7). The first and the second requirement make also reference to some calls concerning advanced technologies for localization, mapping, diagnosis and restoration of the tangible heritage in relation to climate change (as from FP1 to FP6).

Creative Europe and *Erasmus +* are the main programs of the DG EAC for 2014-2020. The first one presents several funding opportunities to administrators that want to engage in projects of cooperation or in the constitution of networks and thematic international platforms concerning the cultural field and the economy of culture. The second one, instead, supports programs of education, training, research and mobility for the actors of the field (from students to public employees).

Other opportunities are offered by the DG ENTR through the COSME program which is focused on internationalization and on the access to credit for small and medium businesses in the fields of fashion, design, craftsmanship, manufacture. Finally, there are the programs of territorial cooperation financed by ERDF and coordinated by the DG REGIO, i.e. INTERREG V, the Program for rural development 2014-2020 and URBACT III.

Table 5. Thematic priorities in Cohesion Policy 2014-2020

EU 2020 Objectives	Cohesion Policy 2014-2020 - Thematic priorities
Smart Growth	1. strengthening research, technological development and innovation (R&I);
	2. enhancing access to and use and quality of information and communication technologies (ICT);
Sustainable Growth	3. enhancing the competitiveness of small and medium-sized enterprises, the agricultural sector (for the EAFRD) and fisheries and aquaculture sector (for the EMFF) (SMEs);
	4. supporting the shift towards a low-carbon economy in all sectors (low-carbon economy);
	5. promoting climate change adaptation, risk prevention and management (climate change);
	6. protecting the environment and promoting resource efficiency (environment);
	7. promoting sustainable transport and removing bottlenecks in key network infrastructures (transport and infrastructures);
Inclusive Growth	8. promoting employment and supporting labour mobility (jobs);
	9. promoting social inclusion and combating poverty (social inclusion);
	10. investing in education, skills and lifelong learning (education);
	11. enhancing institutional capacity and an efficient public administration (public administration).

Source: KEA (2012)

Figure 5. *Culture in the 2014-2020 European Programming*

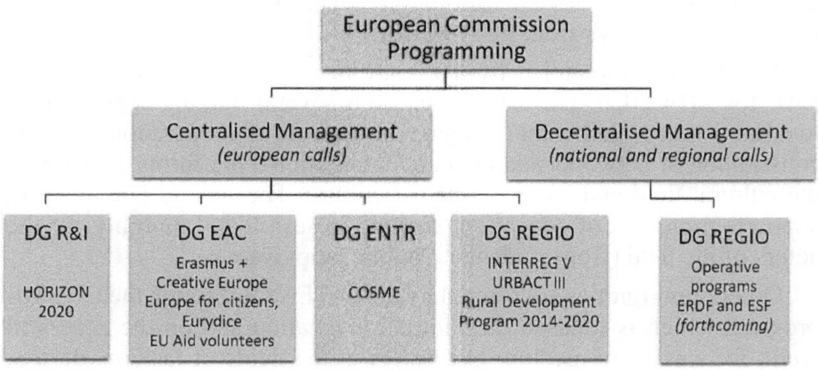

Source: elaboration of the author on Kern and Montalto (2013)

In short, European cultural programming lived its highest moments in 1993 with the introduction of art.128 of the Treaty of Maastricht, in 2007 with the entrance of creativity in high-profile documents and in 2010 with the Europe 2020 strategy, which paves the street for the future. All along the path, the KEA reports are fundamental despite they constantly register a gap between the economic development policies of the Cohesion Policy, still limited to tourism and natural and cultural assets, and those directly handled by the Directorates. Perhaps, this discrepancy is due to the agreements with Member States and Regions for the NOPs and ROPs and the expenditure of ERDF and ESF funds.

In consideration of that, the main organizations of the cultural and creative field, both public and private, joined together to induce the EU and the Member States to include cultural heritage among the EU's fields of action and to adopt a European policy agenda on the theme. In this sense, the research projects, scientific divulgation, information campaigns, created and supported in the last five years by non-governmental organizations are fundamental, e.g. those by Europa Nostra, *Heritage Europe*[46] and ENCATC[47]. For instance, in 2011 Europa Nostra launched the *European Heritage Alliance 3.3*, i.e. a network of thirty organizations supporting cultural heritage recovery and re-use. They worked together for the insertion of cultural heritage among the EU's political priorities[48]. Registering the absence of a complete and exhaustive review of the EU programs and projects on cultural heritage, the alliances first financed a research project that illustrated, both qualitatively and quantitatively, the projects' impact on territories according to a clear set of parameters: quality of life (beauty, harmony, and sustainability of the life in European settlements – both urban and rural); sense of identity and belonging of the European citizens; working opportunities; sustainable and intelligent development of European economy[49].

The project *Cultural Heritage counts for Europe* (CHCfE) was launched in 2013 in order to proof the key role of cultural heritage in the Europe 2020 strategy, foreseeing its intermediate revision (2016). The project produced a reference framework for the study and analysis of cultural heritage policies though the collection and survey of about one-hundred case studies (CHCfE Consortium, 2015, p. 17). The data presented in the project report had great resonance among European institutions. In 2014 the Council of Ministers of Culture adopted the "Conclusions on cultural heritage as a strategic resource for a sustainable Europe" (semester of Greek presidency), and the "Conclusions on participated governance of cultural heritage" (semester of Italian presidency). The European Com-

46 *European Association of Historic Towns and Regions.*

47 *European Network of Cultural Administration Training Centers.*

48 The launch of the network took place within the online project "Mainstreaming Heritage" by Europa Nostra (*europeanheritagealliance.eu*). The name of the website refers to art. 3.3 or the Treaty of Lisbon that says: "[The European Union] respects the richness of its cultural and linguistic identity and ensures its safeguard and the development of the European cultural heritage" (*eur-lex.europa.eu*).

49 See European Heritage Alliance 3.3, 2012. "Towards an EU Strategy for Cultural Heritage – The Case for Research", *position paper* presented to the European Commission in April 2012 [on line] Available on: http://www.europanostra.org/UPLOADS/FILS/Towards-an-EU-Strategy-for-Cultural-Heritage_final.pdf(last access: 19.02.2016).

mission also adopted a "Communication towards an integrated approach for cultural heritage in Europe" followed by the report *Getting Cultural Heritage to Work for Europe* of the Horizon 2020 experts in cultural heritage and by the Conclusions of the European Council on the *Work Plan* for Culture 2015-2018[50].

The report *Cultural Heritage to Work for Europe* contains recommendations for the definition of an innovative framework for public policies and research on cultural assets and innovation up to 2020, while the Conclusions of the Council identify cultural heritage as one of the four priorities of the plan for culture 2015-2018. Thus, they confirm the necessity for the European Union to invest in cultural statistics, as a preliminary condition for the elaboration of evidence-based policies for cultural heritage (CHCfE Consortium, 2015, pp. 9-11).

2.4 Conclusions

In the international charters and declarations, the connection between cultural heritage, creativity and innovation is now as factual as the link between the theories on creative systems (clusters and networks) and those on cultural and creative industries. This emphasizes the role of invention in the cultural heritage chain and in the various phases: safeguard, knowledge and diagnosis, protection and preservation, valorization and fruition, management (Amari, 2006; Santagata, 2007; Mautone, 2009; vanOers, 2010).

Cultural heritage, conceived as a set of cultural and natural assets, including landscapes, constitutes a complex system. The management of this system requires a clear and transparent frame of reference. For instance, the introduction of culture and creativity within European documents showed how the EU programs and projects on cultural heritage were fragmented and incomplete, therefore pushing the EU authorities to adopt a specific policy for culture. The first initiative in this was the "bottom-up" project *Cultural Heritage Counts for Europe* which researched and classified all the project (completed or not) funded by EU since 1980 (CHCfE Consortium, 2015).

In summary, the creation of a clear and transparent knowledge network for culture and creativity still remains the main challenge in policy making and research on culturale heritage and landscape.

50 For further information see http://ec.europa.eu/culture/policy/culture-policies/cultural-heritage_en.htm (last access: 19.02.16).

Creativity at the Regional and Urban Scale

High profile documents and programs of international organisms, such as UNESCO and the European Union, are put in practice on a local scale through policies and strategies created by national governments according to the principle of multi-level governance. This chapter investigates how local authorities translate these documents into practice at the regional and urban scale. Above all, the chapter focuses on the taxonomies and narratives adopted in describing the cultural and creative clusters/networks, the purpose of their intervention (urban renovation, entrepreneurship, social inclusion, international cooperation, institutional capacity building) and their spatial consequences with respect to the historical urban landscape (recovery, requalification, new buildings).

3.1 The international context

Asia

Cultural and creative networks always operate towards a requalification of the urban system, taking as a reference the British financial tools and the American *modus operandi.* The large-scale urban plans are clearly thought to support a local economic agenda and, at the same time, promote the image of a globally relevant city.

With this respect, it is possible to remember the *West Kowloon Cultural District development plan* and the *Jockey Club Creative Arts Centre* in Hong Kong, the *Hong-dae Cultural district* and the *Itaewon* neighborhood in Seoul, the "creative industry parks" in Beijing and Shanghai, born in occasion of the Universal Exposition of 2008 and 2010. In general, the strategies for culture of Asian cities lacks a space of confrontation where to share, discuss, and debate ideas. This provides no foundation to these policies, raising once again the question on how to establish global policies within the everyday life of cities (Hutton, 2016; Kim, 2015; Kim, 2015; Stevenson, 2014; Tang, 2015).

Oceania

Australia

The development of cultural and creative networks in Australia is strongly connected to the international activity of scholars and professionals from the United Kingdom. From the 1980s they "exported" the British model of policies for the arts and artistic communities (the so-called planning for the arts), adapting it to the different contexts and proposing it to local governments. In the 1990s, the focus on artistic production, in the broadest sense of the word[1], became a peculiar feature of the Australian cultural programming. It managed to influence even most recent phenomena, such as the multiplication of temporary events and spaces (spontaneous or inserted in more articulated and complex strategies, as big events) or the redaction of strategies supporting creative economy (in which the words culture and creativity are often used interchangeably in the description of the actions in support of the arts).

In relation to the first point, the initiative *Renew New Castle* promoted by Marcus Westbury in the homonymous town of *New South Wales.* On the other hand, as far as creativity is concerned, it is possible to remember the city of Brisbane, where the urban strategy for culture, initially defined as Cultural Development Strategy (1991),was renamed by the City Council *Creative city: Brisbane Cultural Strategy* (2003-2008), even though the interventions were only partially concentrated on creative networks and their spatial dynamics. On this behalf, interesting are the studies conducted on the phenomenon of *clustering* in the cultural and creative neighborhoods of Fitzroy/Collingwood in Melbourne and Surry Hills in Sydney (Roodhouse, 2010; Stevenson, 2014; Wood and Dovey, 2015).

North America

Canada

The Canadian initiatives in support of cultural and creative networks are affected by a British and American influence, as far as content and purposes are concerned, above all for the accent on a community and neighbor dimension. Yet, they differ from them for a better integration within municipal planning. It is possible to remember, for instance the *Toronto Culture Plan,* the *Ottawa Arts Plan* and the recovery of Yaletown in Vancouver. Unlike in the United States, interventions are not limited to the requalification of the building or the neighborhood, influencing different spaces and facilities of the urban ecosystem to answer the client and the final consumers' needs. This has been proved by projects as, for instance, the *Niagara Culture Plan*, the private initiative *Toronto Artscape* and the program *Arts Habitat Edmonton* (Duxbury, 2004; Hutton, 2016; Marontate and Murray, 2010; Stevenson, 2014).

1 Anthropologic definition of culture (culture = lifestyle of a community).

United States

The United States approach towards cultural and creative networks is based on a strong philanthropic tradition in the field of the arts, which favored a business-oriented approach to culture and more recently, towards creativity. The urban and regional strategies that derived from it are strongly aimed at economic and social development and are mainly manifested through the re-development of buildings and neighborhoods. This was the dominant approach in the case of the *Our Town Scheme* of the *National Endowment for the Arts* (NEA), of the Strategic Plan of *S. Louis 2004* or in the projects that concerned urban brownfield: Mid-Manhattan in New York, Wicker Park in Chicago, South of Market Area in San Francisco. Nonetheless, these projects reflect an interventionist approach based on physical transformation and economic incentives[2] as the attractive element for creative and cultural networks and their settling in the territory (often, with disappointing outcomes in the long run). Few are the cases where economic development took place thanks to the only circulation of information, in absence of a public plan or leading cultural experiences. Among these, Austin is for sure to include (Borrup, 2015, Corò and Della Torre, 2007; Frost-Kump, 1998; Hutton, 2016; Murdoch *et al.*, 2016; Noonan, 2013; Roodhouse, 2010; Rushton, 2015; Sacco and Pedrini, 2003; Sacco and Ferilli, 2006; Stevenson, 2014).

3.2 The European context

Austria

In the Austrian context, the better-known case is that of Linz. The transformation from city of steel production to city of culture sees the Plan of Cultural Development of 2000 and the city's nomination as European Capital of Culture 2009 as two main references. These two initiatives made possible the reconversion for cultural use of several degraded spaces of production but also the reconversion of the local productive system thanks to the integration and the cooperation between the artistic-cultural sector (computer art) and the scientific-technological one. A union that gave Linz the UNESCO award *City of Media Arts* for 2016 (Corò and Della Torre, 2007; Roodhouse, 2010; Sacco and Ferilli, 2006; Sacco and Pedrini, 2003).

2 In this case, the tax of purpose introduced in Denver to finance the public agency *Scientific and Cultural Facilities District* (SCFD) is emblematic (Sacco and Ferilli, 2006; Sacco and Pedrini, 2003).

France

The French experience defines itself for a dual approach to the creative and cultural networks provided by: 1) interventions on neighborhoods in metropolitan areas; 2) interventions on a district scale in suburban/regional areas. The first category corresponds to artistic and cultural projects accomplished in the neighborhoods of Paris, Lyon, Marseille, Montpellier, Rennes, while the second category refers to the productive and touristic districts of the Loire Valley and the *Pays de Valois* (France). The French model proposes an alternative solution to cultural and creative industries of the British DCMS. This is defined as "culture-centric" and is characterized by the integration between projects on "high-culture" carried out by a limited number of important national cultural institutions and projects realized closely with local organizations, professionals and amateurs, following the anthropological concept of culture. (Andreas *et al.*, 2012; Bianchini and Parkinson, 1993; Corò and Della Torre, 2007; Montgomery, 2003; Sacco and Ferilli, 2006; Sacco and Pedrini, 2003; Usai, 2013).

Germany

Cultural and creative networks became increasingly strong and important after national unification, contributing significantly to the requalification of large abandoned industrial areas (the so-called "*creative dispossessions*"). This produced cultural neighborhoods that shortly after their requalification became relevant centers for contemporary art and even reference centers for urban life. Among these, it is possible to count the *Adershof District* in Berlin, the *Zolleverein Area* in Essen and *Baumwollspinnerei* in Leipzig. In some centers, the creative dispossessions were oriented toward technological innovation, thanks to the support of local universities, as for Dortmund technological park. Detached by its industrial past, it brought to the definition of the Ruhr area as the "*Learning city*". The process of reconversion of the German productive tissue contributed to the definition of local and national identity and is therefore recalled in literature, to prove the possibility of supporting cultural and creative economy without foreign-born models (Alfken *et al.*, 2015; Bianchini and Parkinson, 1993; Ferilli at al., 2016; Hutton, 2016; Sacco, Ferilli, 2006).

Norway

In Norway, as in the other Northern countries[3], the national policies on cultural and creative networks represented a valid framework for regional and urban planning, instead of constituting a common strategy of reference. In fact, the initiatives applied to a local scale differ from each other

3 For Sweden, see the case of Stockholm in Sacco and Ferilli (2006).

and concentrate on different sectors of cultural and creative economy. The case of the city of Kristiansand, in the Southern region of Adger, is emblematic. In fact, within the multi-sector regional program *Creative Energy*, focused on art, culture and education investing in the construction of the *Kilden Performing Art Center* (*Lille Cultiva* project). Bergen, instead, built its reputation of cultural and creative city from its role of cultural heritage center (European Capital of Culture in 2000), in order to later become an industrial center and a welcoming city (*Action Plan for the Creative Industries* 2008-2011). The Norwegian experience poses the problem of integration of the international definitions of creative and cultural industries on a local scale, not to mention the integration between a top-down and a bottom-up approach in this process of adaptation (Mercer, 2008; Pinheiro and Hauge, 2014; Sjøholt, 1999).

Nederland

After the bombing of Second World War, several cities of the Nederland had to rethink their spatial configuration. The urban strategies of those years supported, long before the rise of creative cities, local cultural and artistic networks, which developed with no continuity and later acquired a central role in the interventions of requalification in public spaces. These interventions took as their cultural reference the American model with no critical thinking, as in Rotterdam with the requalification of the *waterfront*, inspired to the interventions of Toronto, Baltimore and Boston and, as in Amsterdam, with cultural policies openly inspired to Florida's theories. This produced unsuitable design solutions for the cultural landscape of Dutch cities (historically very dynamic and differentiated), incentivized by the use of resources destined to the interventions of social inclusion for the constitution and the strengthening of a new middle class (creative class and creative artists). The application at the local scale of the international definitions of creative and cultural industry has been found problematic in the Nederland as well (Bianchini e Parkinson, 1993; Peck, 2012).

Portugal

It is not easy to find studies on the Portuguese creative-cultural field in scientific literature. On the one side, this is due to the difficulty in finding tools and methods of survey capable of describing agglomerative phenomena of cultural and creative networks. On the other hand, this also depends on the scarce interest that clustering raises in the local system of research, more interested in qualitative aspects (segments of cultural production, analysis of the policies adopted on various levels, spatial distribution of the enterprises of the field). On this behalf, recent research (to overcome the gap of information) confirm the metropolitan dimension of the cultural and creative Portuguese networks. These are above all concentrated in the "most appealing" areas of Porto and Lisbon, and the relevance of big events: European Capital of Culture (Lisbon 1994, Porto 2001), Portuguese Capital of Culture (Coimbra, Faro), and Universal Exhibition (Lisbon 1998) (Costa, 2007, 2015).

United Kingdom

The United Kingdom is a precursor in the field of policies for the creative and cultural field. The first initiatives can be traced back to the end of the 1980s, when some avant-garde studies[4] already signaled the need of an enlarged approach on public policies for culture. The *Greater London Council,* in their wake, managed to action the first urban strategies in support of collective arts and artistic communities of London's neighborhoods: from the requalification of *Tate Gallery* and London Docks, to the interventions in Soho, Clerkenwell and Shoreditch (Hutton, 2016; Stevenson, 2014). On the London example, district experiences rapidly spread all around the country, often with the support of big events, such as the European Capitals of Culture: Birmingham, Glasgow, Liverpool, Newcastle, Manchester, and Sheffield. The approach adopted in these initiatives is the one delineated by the Department of Culture, Media, and Sport (DCMS). It has been defined in literature as "econo-centric" for its focus on entrepreneurship, private initiative and on the creation of specific governing agencies to compete with the private actors (Andreas *et al.*, 2012; Bianchini *et al.*, 1988; Comunian and Sacco, 2006; De Propris and Lazzeretti, 2009; Evans, 2001; Mould and Comunian, 2015; Roodhouse, 2010; Sacco and Ferilli, 2006; Santagata, 2002; Stevenson, 2014; Usai, 2013).

Spain

The creation and the support to creative and cultural networks in Spain is based on the cultural and artistic urban requalification of strategic neighborhoods within the main cities, as in Madrid, Barcelona, Bilbao, and Valencia.

This model of intervention uses abundantly public finance and public-private partnership, immediately guaranteeing great results (comparable to the big events ones). Yet, it shows serious management issues in the long run, showing problems in maintaining the same performance of the beginning (reorganization and repurposing of cultural facilities, once established the number of users). In addition, it produces gentrification of the recovered areas (stagnation of the creative class, when urban profits become too high, conflicts with locals and cultural enclaves) (Bianchini and Parkinson, 1993;

4 We are referring to Franco Bianchini, Mark Fisher, John Montgomery, and Ken Workpole's studies.

Martí-Costa and Pradel i Miquel, 2011; Evans, 2001; Pareja-Eastaway and Pradel i Miquel, 2015; Sacco and Ferilli, 2006; Sacco and Pedrini, 2003).

3.3 The Italian context: the national dimension

3.3.1 Cultural and creative economy in Italy (1990-2009)

In Italy, the birth of new production chains connected to urban renewal and to the valorization of cultural heritage, in spite of the crisis of traditional industry, pushed public administrations to engage in creative spatial policies.

Starting from the early 1990s, local administrators tried to quantify the cultural field and its added value[5]. The 2004 Bodo and Spada's report on culture economy was the first attempt to quantify the cultural market in Italy and its added value for the period 1990- 2000 on the basis of the public and private spending[6] and national accountancy coefficients (MiBAC and IstitutoTagliacarne, 2007, pp. 17-21).

In 2001 the reform of Title V of the Constitution redefined the responsibilities by the State and the Regions with respect to the cultural heritage. Heritage protection was assigned to the former while heritage valorization to the latter.

Heritage valorization and cultural services, already subject to the regulations on public works, became a complex matter requiring new operational tools. Between 2000 and 2003, several normative interventions tried to handle the question comparing it to the managerial issue posed by the programs for infrastructures and urban renewal (Table 6). However, the strategies of national governments, mainly aimed at the privatization and alienation of the public historical properties, were fruitless. In this period, the production of reports on cultural economy stopped (Colavitti and Usai, 2014; De Medici,2010; De Medici and Pinto, 2012; Fabietti, 2008; Ponzini, 2008; Tartaglia, 2006).

In 2004, the Italian Code on cultural heritage and landscape introduces further forms of valorization to renovate the public cultural planning and its governance[7]. Cultural heritage and landscape are fundamental components

---

5 See Centro Studi Touring Club Italiano, *Indagine sulla consistenza del patrimonio culturale immobile di interesse turistico*, July 2003 (survey conducted for MiBAC).

6 Evaluation of State, Regions, Provinces, and Municipalities' spending and of the expenditure of families for advertisement, sponsorships and donations.

7 The DLgs.42/2004 defines cultural assets as inalienable public goods. Their property can be transferred only to other public administrations (art. 10). The alienation of a cultural asset is possible only if the verification of authenticity has a negative result or if the MiBAC previously provides an authorization certifying the compatibility of reuse project (art. 54-55). In addition to this, the Code defines the forms of management and valorization for cultural heritage. They are divided into direct and indirect forms (artt.112, 115, 116). The latter are the concession to third parties or the administration

of the country along with the system of cultural and creative enterprises working in the field, as emerged in the survey *La città dei creativi* (The city of the 'creatives'), published in 2000.

The research realized by the *Associazione Nazionale dei Costruttori Edili* (ANCE) (National Association of Building Constructors) promoted the quality of the environment and the leadership in urban areas as fundamental conditions for the attraction of creative talents (ANCE and Ambrosetti, 2005).

Table 6. Legislative interventions concerning the public assets of historical-cultural value

L.D. 490/1999 art. 54	It takes over the L. 127/1997 establishing that the State's properties are to be considered non-sealable public goods. However, according to art. 24 of the L.1089/1939, it establishes that alienations are possible with a prior Ministerial authorization (which includes a program of valorization for the good alienated).
L.144/1999	Project Financing: institution of the technical financial Unit of the project
L.488/1999	Modifies the previous norms and fixes new rules for the transfer of property rights on public heritage assets and for the concession and use of the same by private through convention
L.342/2000 art. 38	Provisions for the development of institutional tasks and for the realization of cultural programs within the field of cultural heritage and entertainment.
L.388/2000 art.145 comma 97	The Ministry of Public Works, along with the Ministry of Economy can authorize highways and tunnels franchises to invest in the recovery of monuments, building, and artifact of historical-artistic value, and the valorization of the city.
D.P.R. 283/2000	Decree of implementation. It allows the Ministry of Economy along with MiBAC to alienate State-owned buildings of historical and artistic value waiving the art.54 of the L.D.490/99. It fixes the conditions (MiBAC authorization, pre-emptive right for the State in expensive alienations, elimination of alienation rights for the violation of procedural norms). Definition of the inalienable assets.
L. 410/2001	Privatization of public heritage and real estate funds (securitization, concession of valorization). It creates the *Società cartolarizzazione immobili pubblici* (SCIP Ltd), a vehicle company that receives the property to be alienated by the State and other public organizations, in order to realize it securitization.

through subjects constituted and/or participated prevailingly by the public administration which holds the asset (e.g. institutions, foundations, associations, consortia, limited companies, etc.). On this topic, see Colavitti and Usai (2013a, 2013b, 2014).

L.448/2001	It allows the concession to private subjects of the cultural services in public museums and sites.
Decree of the Ministry of Culture (MiBAC) n.491/2001	Decree on the constitution of/participation in private foundations by MiBAC.
L.112/2002	Institution of *Patrimonio* Ltd. for the valorization, management and alienation of public assets (among which there are potentially also non-transferable cultural goods). *Patrimonio* Ltd. has to finance *Infrastrutture* Ltd.
L.137/2002	It articulates and rules the public-private partnerships, including the mixed foundation.
L.289/2002 art. 60	3 % of the State allocations for infrastructures are destined to protection and recovery of cultural heritage and for cultural activities.
Decree of the Ministry of Infrastructures and Transport (MIT) 03/05/2002	At the Ministry of Infrastructures and Transport is set up the Commission infrastructures and culture. The commission works on accessibility to cultural and natural heritage sites.
L.291/2003 art. 2	Institution of ARCUS Ltd. at MIBAC to finance the conservation of cultural and natural heritage and cultural activities.
L.326/2003	It provides the mechanism of silent-assent in the verification of no-conflict of interest for State property that the government wants to alienate.
L.326/2003	Institution of the Urban Development Companies (UDCs) for the valorization of the State's cultural and natural heritage.
L.D. 173/2003	The State Property Office becomes a public economic organization with a modification of the art. 65 L.D.300/1999.

Source: elaboration of the author

The ANCE came to this conclusions earlier than other institutional players, which started getting interested in the subject, after the L.77/2006 for the introduction of the UNESCO management plans in Italian sites[8].

The L.77/2006 represents a crucial moment for the cultural and creative sector in Italy, under the profile of valorization of public buildings of

[8] Italian good practices in the elaboration of the UNESCO Management Plans are collected in Ferroni and Patrigani (2013).

historical-cultural interest because it ended their alienation in favor of the concession in use to privates for a long period[9].

In 2006, the report realized by KEA for the European Commission introduced a methodology for the quantification of the economy of culture. It focused on the "added value" deriving from the expenses for the cultural and creative field (Table 7). The distinction between the cultural and creative field by KEA made Italian governments aware of the preponderance of heritage conservation in respect to the production of "new culture".

Table 7. The cultural and creative field in the KEA report (2006)

CIRCLES	SECTORS	SUB- SECTORS
CORE ARTS FIELD	Visual arts	Crafts Paintings - Sculpture - Photography
	Performing arts	Theatre - Dance – Circus - Festivals
	Heritage	Museums - Libraries - Archaeological sites - Archives
CULTURAL INDUSTRIES	Film and Video	
	Television and radio	
	Video games	
	Music	
	Books and press	
CREATIVE INDUSTRIES AND ACTIVITIES	Design	Fashion design, graphic design, interior design, product design
	Architecture	
	Advertising	
RELATED INDUSTRIES	PC manufacturers,MP3 player,manufacturers, mobile industry, etc…	

: "the cultural sector" (grey)
: "the creative sector" (black)

Source: MiBAC and Istituto Tagliacarne (2007)

9 From a normative point of view, the references are the L.D.156/2006, the L.D.62/2008 and the L.113/2008.

In 2007 Walter Santagata published the book *La fabbrica della cultura.* The author stated that a country is culturally productive when the cultural production exceeds heritage preservation. In order to make this possible, Santagata (2007) proposed to the production chain of the cultural assets according to the studies on innovation and, above all, to the systemic theory of the neo-Schumpeterian school. He proposed a model of production chain, the *cultural heritage chain*, constituted by the following phases: selection of the artists, creations of the ideas, production, distribution, and consumption. Most of statistic studies on cultural economy produced in Italy in the next three years referred to this phases. For instance, the report realized by the Istituto Tagliacarne for MiBAC in 2007, according to the 5-digit national statistic classification ATECO 2020[10]. Analyzing the international statistic models for quantification of cultural and creative economy, the survey individuated a "common nucleon" of fields, formed by cultural industry and by the heritage (assets and cultural activities). It put aside the macro-sectors of tourism and of transportations and concentrates on "cultural" activities, adopting a classification that constituted the methodological base for the following MiBAC's White Paper. To these cultural activities, strictly connected to the heritage (18 categories), the activities connected to the territory are added: certified products and productions in wine and food industry (22 categories), the productions of industrial and artisanal nature (43 categories), urban renewal and architecture. These latter were a recurring theme in international models but, in the Italian one, they were enriched by a new value connected to the use of innovative materials, recovery and re-use of public heritage (MiBAC and IstitutoTagliacarne, 2007). The methodology developed by IstitutoTagliacarne was adopted in a statistic survey on the use of technologies for heritage supported by the MiBAC's General Direction for Innovation[11] and by Enea[12].

The data so collected were then used for an on-field research by Unioncamere whose results were transposed in the report *Il sistema economico integrato dei beni culturali*, published in 2009. The report tried to prove the existence of a cultural heritage chain in Italy, in close relation with the territory and its cultural offspring. The study was structured on

10 The Italian Classifications of economic activities (ATECO) adopted by Istat in 2002 was substituted by the 2007 ATECO classification rounded to the sixth significant digit and compatible with the NACE European classification rev.2 rounded to the fourth significant digit and with the UN to the third digit (*International Standard Industrial Classification of All Economic Activities, Rev.4*) (ISTAT, 2009; MiBAC and IstitutoTagliacarne, 2007 p. 24).

11 During which a sample of companies of the field filled a questionnaire on the subject.

12 Cfr. ENEA – Ente per le Nuove tecnologie, l'Energia e l'Ambiente (Organization for New Technologies, Energy, and Environment) (2007). *ENEA e le tecnologie per il patrimonio culturale*, Dossier of the Workshop "Distretti e tecnologie per i beni culturali" (Districts and Technologies for heritage), Rome 24th Ottobre 2007, Frascati: ENEA.

the sequence of knowledge, protection, management and valorization proposed by Walter Santagata[13] (2007). Furthermore, the report characterized the companies operating in the different phases of the cultural heritage chain through a series of interviews[14] (Table 8) (Unioncamere and IstitutoTagliacarne, 2009).

Table 8. Distribution of the interviews in the different phases of the cultural heritage chain

	Number of enterprises
Knowledge Preservation, control and restoration of the artifacts	0*
Preservation Meta-cataloguing of the asset. It includes traditional cataloguing of all the references to scientific analysis, studies, spatial and temporal collocations, virtual images and tools to guarantee the "life" of the artwork in the "virtual" world.	4
Fruition Modalities of fruition and diffusion, new narrative formats, recovery of buildings and places of historical value, new business models for preservation, safety and sustainable management of heritage sites.	6
Economic management Activities related to the valorization of cultural heritage through tourism for landscape, food and wine, and business.	5
IT platforms Creation of software which accompany the enterprise in all the phases of production	5

**Scientific enterprises or research organizations are mainly small or individual.*

Source: elaboration of the author on Unioncamere and Istituto Tagliacarne (2009)

13 The report individuates 138 categories of the 2007 ATECO classification with the addition of an agricultural field concerning quality products, which cannot be identified with one ATECO activity (ISTAT, 2009). Starting from these activities it is possible to quantify the occupation and the value added produced by the cultural chain, adopting a bottom-up procedure, based on the Italian division of the country in Provinces. As far as occupation is concerned, the reconstruction has been carried out based on the provinces per sector of economic activity and number of professionals, the renewed to total occupation follows a top-down path (starting from national, regional, and provincial evaluations provided by Istat, with the distinction between regular and irregular component).

14 The interviews are preceded by an analysis of the company they belong to, its history, and the competitive dynamics of the reference market.

The sample taken into consideration was not statistically relevant but provided useful elements for the definition of the companies working in the Italian cultural heritage chain.

The subject has been taken over and broadened in the *Libro Bianco sulla Creatività* (The White Paper on Creativity), edited by the MiBAC Commission of creativity and culture production in Italy, chaired by Walter Santagata. The White Paper had the merit to apply the chain analysis to cultural and creative economy. Thus, it made the studies on the added-value of cultural and creative economy and those on the cultural heritage chain converge overcoming the traditional statistics (Table 9 and Figure 6) (MiBAC, 2009; Santagata, 2009).

Table 9. The cultural and creative sector in the White Paper

Industrial production	Fashion Industrial Design and Handicraft Food and Wine Industry
Cultural goods and services and Communication	Software Publishing TV and Radio Advertising Film industry
Historical and artistic heritage	Cultural heritage Music and live performances Architecture Contemporary Art

Source: MiBAC (2009)

Figure 6. *Contribution of the cultural and creative field to the national economy in the different phases of the cultural chain*

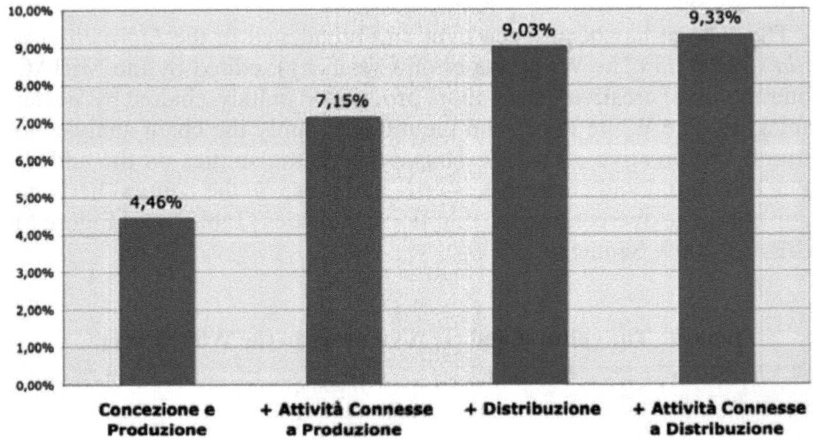

Contribution of the cultural and creative economy to the national GDP (%)
Invention and production
Satellite activities of production
Distribution
Satellite activities of distribution
Source: MiBAC (2009)

Within the international landscape, the White Paper identified two approaches to creativity. The first one was connected to the technological innovation and it was characteristic of North-European and North-American countries. The second was related to social equality[15] and it was typical of the Mediterranean area and of African, Latin American and Asian countries. According to this report, Italy represented a conjuncture between the two models, since it read creativity through its territory, putting together the technological progress and the ethic and social one[16]. However, there were still

15 The White Paper defines social equality as the measure in which people are able to participate to social, economic, and cultural life, and to the development of their communities in favorable conditions for individual well-being and potential. What allows individuals to access culture and cultural heritage in its different forms, access fundamental goods (good health, increase human capital) or complex goods (participate to community life, improve the projects) of the city.

16 The core of the Italian model is the connection between creativity and culture production with history and territory. Time and space are historical dimensions of extraordinary waves of creativity that accompanied the history of the Country. Cultural and creativity industries are strongly marked by the territory, which, in the Italian case, means above all production of material cultures, urban spaces and economic districts (MiBAC, 2009, p. XV).

several obstacles to overcome: an inefficient management, a reduced demand, a lack of economies of scope (for instance, cultural tourism). In this view, the White Paper supported the development of adequate scale and agglomeration economies as a pre-condition to creative spatial policies, this through the relational networks spread in the city and over the territory (Table 10).

In the field of architecture the White Paper proposed the following strategic actions: the candidacy of an Italian modern architecture as UNESCO site; measures for the internationalization of Italian architects; development of the research and development departments in local firms and in the education centers. As far as cultural heritage is concerned, the report supported: volunteering and free access; the definition of standards of quality; the extension of the museums services; tax benefits on private donations. Lastly, within the field of live music and entertainment, the White Paper proposes the creation of a fund for creative capital, supported by MiBAC, by the Ministry for Economy, and by the Ministry of Education, Youth, and Sports, with the largest banks of the country (Action 70). The fund was supposed to destine an annual sum, determined according to availability, for artistic and creative projects on the base of a rigorous peer review process, conducted by international experts, with shares reserved for particular targets (young artists, artists with a meaningful career, non-profit associations, etc.). Action 70 is important since it was one of the few to be implemented. Thanks to the effort of the *Giovani Artisti Italiani* (GAI, Young Italian Artists) and the Ministry for Youth the project *Italia Creativa* was established.

Table 10. Strategic options for the development of the economies of scale and agglomeration in Italy

Creative cities and neighborhoods	– "National" policies aimed to and integrated with local policies; – A *Creative Social Network* or *database* of professions and creative industries; – Enhancement of cultural heritage and historical centers through a special program for the city centers aimed at their livability, functionality and connection with the surroundings and an increase in the sales of additional services and new forms of sponsorship and patronage. – Intervention on the urban plans and zoning programs, which help dedicate some urban areas to people from different social classes and mixed uses.
Industrial and touristic districts or clusters	– If the district is artificial (created by institutions), the aim of the public field is to "naturalize" it, i.e. create the right conditions for cultural enterprises to act autonomously in the invention phase.

Source: elaboration of the author on MiBAC (2009)

The *Italia Creativa* project supported and promoted the young Italian artists through networking activities that took place on the whole national territory from the leader cities of the GAI network, in collaboration with other local stakeholders. The report *Economia della cultura e giovani* offers a complete list of the initiatives realized in the cities involved in the project. The survey also provided socio-economic data for each city identifying the public actions for the birth of new creative systems. The relevant factor is that, in spite of a five-years distance from the ANCE report, both studies underline the importance of urban renewal in the construction of creative cities and neighborhoods (Cicerchia, 2010, 2013). Lastly, the GAI project marked the entrance of Federculture[17] in the Italia statistic reports on the creative and cultural economy, with the proposition of a first Italian index of creativity based on production and not on consumption, as in Richard Florida and Irene Tinagli (2005).

3.3.2 The creative capacity of culture in time of austerity (2010-2014)

From 2009, the economic crisis and the limitations imposed to public expenditure, accelerated the diffusion of market-oriented management in the recovery and valorization of public cultural heritage, with the diffusion of public-private partnerships (Agenzia del Demanio, 2012; DeMarchi, 2009; Gambaro, 2006; Tartaglia, 2006). The procedural and normative mechanisms developed for the *Società di Trasformazione Urbana* (STUs, Urban Development Companies) and project financing were extended to the dispossession of State properties as well as in the National Plan for the City of 2011[18] (Agenzia del Demanio, 2012; ANCI, 2010; Antoniol,

17 Federculture is the Federation of the Companies and of the Organizations of management of culture, tourism, sport, and leisure time. This association, born in 1997, represents the most important cultural enterprises of the Country, together with regions, provinces, municipalities, and other public and private subjects involved in the management of the services connected to culture, tourism, and leisure time (www.federculture.it).

18 State's property federalism (italian: *federalismo demaniale*), consists in the attribution of the ownership of some state buildings to municipalities, provinces, metropolitan cities, and regions in order to provide those local administrations with a heritage of their own, as in L.42/2009 on fiscal federalism. This is possible by making "transferrable" unavailable State's property and assets, meaning to favor their passage to the available heritage with the possibility for local administrations to alienate those goods, confer them in real estate funds, manage them directly or give them to third parties (L.D. 85/2010). The National Plan for the City, on the other hand, finances coordinated interventions of re-use and regeneration of degraded areas on the whole national territory (L.134/2012 art.12, MIT Decree n.286/02012 and n.349-2012). The procedure implies that the Municipalities present their proposal to the *Associazione Nazionale Comuni Italiani* (ANCI, National Association of Italian Municipalities), through an online one-stop-point. The proposals collected by ANCI are selected by the MIT'S control booth. On the base of the presented project, the administrations involved sign a contract of

2010; Federalismo.sspa, 2011; Storto, 2013; Zerman, 2012). In this process, the buildings of historical-cultural value had such a relevant role that they required new instruments of valorization for an efficiency project cycle management: the *Piani strategici di valorizzazione e sviluppo culturale* (PUVAT, Strategic plans of valorization and cultural development), the *Contratto di Valorizzazione Urbana* (CdV, Contract of Urban Valorization)[19] (De Medici, 2010; Ponzini, 2008; Giotta and Patruno, 2012).

Italian cities were asked to rationalize the spending and maximize their results concentrating on a small number of strategic interventions according to: the potentials of the social and economic tissue of which they are part; the alliances that they allow to create; the multiplicative mechanism they trigger.

Starting from literature on clusters and networks, different empirical studies tried to describe the connection between territory and creative systems, to analyze the advantages and the problems coming from it, above all in the urban landscape, in order to understand how to handle public resources and what policies to activate.

The report of the European Observatory of the Clusters on territorial concentration of cultural and creative industries on a regional basis is emblematic (Unioncamere and Symbola, 2011; Power, 2010). It underlined how the cultural and creative sector registered different performance according to the regional levels of employment and competitiveness, specialization and clustering (more than to the job market and the number of inhabitants). In addition to this, the analysis registered the highest rates of economic wealth in the regions with the greatest concentration of cultural and creative industries. At national level, the 4th National Conference of the Assessors to Culture and Tourism, organized by Federculture, produced a document of reforms. It suggested a rationalization of public expendi-

---

valorization, which regulates the schedule of the various public and private subjects, in relation to the area to be ameliorated, in order to allow works to start on a defined time (ANCI, 2012). The approach to State's property federalism and of the National Plan for City clearly look up to the *Project Cycle Management*, implying a mechanism of multiplication for the funds. On this topic, see Colavitti and Usai (2013a, 2013b, 2014).

19 The *Programmi Unitari di Valorizzazione Territoriale* (PUVAT, Integrated Programs of Territorial Valorization) were introduced by the L. 214/2011 for the recovery and re-use of public buildings. They are put in practice by agreements of valorization between State, Regions, local governments and through the constitution of new juridical subjects. The PUVAT has been used in the alienation or concession to privates of State-owned historic buildings to municipalities and provinces. In Italian literature this process of dispossession is known as *federalismo demaniale culturale* (L.D. 85 / 2010 article 5 paragraph 5 and Circular MIBAC 18/05/2011). Furthermore, in the National Plan for the Cities were provided bonuses for the Contracts of valorization affecting building dispossessed within the *federalismo demaniale culturale* (ANCI, 2012; Colavitti and Usai 2013a, 2013b, 2014).

ture with a major involvement of the private sector and the constitution of synergies on a local level (Grossi, 2011). The contribution of the city was considered so fundamental which influenced the EU policy documents leading to the adoption of a European Urban Agenda and to its insertion in the framework of Cohesion Policy in the respect of the Strategy Europe 2020[20] (Camagni, 2012).

In the national field the indications of the new community strategy are included through the strategic option[21] "City" of the document *Metodi e obiettivi per un uso efficace dei fondi comunitari 2014-2020* (Methods and objectives for an effective use of European funds 2014-2020), presented by the Minister for Territorial Cohesion on 27 December 2012.

In case of metropolitan areas, the Italian policy documents were aimed at the overcoming of the self-evident contradiction between metropolitan and administrative borders due to missing reforms in public administration and town planning (law of government of the territory) (De Luca, 2012).

On this behalf, the 2011 report *L'Italia che verrà. Industria culturale, made in Italy e territori* (The Italy that will be. Cultural industry, made in Italy and territories) realized by Istituto Tagliacarne for Unioncamere and Fondazione Symbola, proposed a new method for the quantification of cultural and creative economy, which abandoned the Santagata's cultural heritage chain and concentrated on the integration on the different chains on a territory in relation to the activity of cultural and creative local networks. This analytic model, defined as the "cultural platforms model", finds its foundation on the studies on the advanced cultural district. Here culture gets out a specific economic sector (i.e. cultural and creative industries) acquiring a fundamental value in any other form of production

---

20 Here the high-profile documents that were useful for the survey: the Scheme of Development of the European Space (1999); the Third Report on the Commission's Cohesion *A New Partnership for Cohesion* (2004); the document *The territorial state and perspective of the European Union* (2005); the communication of the European Commission n.385 of 2006 on the theme "Cohesion Policies and the city: the contribution of urban areas to the growth and the occupation in the region."; The Green Book on Territorial Cohesion *Turning territorial diversity into strength,* issued by the European Commission (2008); Leipzig Charter (2007); The 2007 Decision of the European Commission n.5313 on *European Observation Network for Territorial Development and Cohesion* – ESPON; the new Treaty of Lisbon (2009); the Declaration of Toledo (2010), the 2010 Communication of the European Commission n.2020 concerning *Europe 2020 – A strategy for an intelligent, sustainable, and inclusive growth.* Finally, the Resolution of the European Parliament of 23 June 2011.

21 The chosen strategic options are: (1) redesign and modernize urban services for the city's residents and users; (2) develop practices of design for social inclusion, to favor the weakest parts of the population and the disadvantaged areas and neighborhoods.

and generating new intangible abilities (creative spillovers) (Sacco *et al.*, 2013a, 2013b; Unioncamere and Symbola, 2011).

To re-launch the Italian model of cultural-based development, in 2012 Unioncamere and Symbola provided in their report some policy-options along with the usual statistics. They gave attention to the re-use of spaces offered by the historical-artistic heritage in an innovative way and not simply as a showroom of past cultural identity. In particular, the report proposed the recovery of remarkable buildings with the following functions: incubators of creative entrepreneurship, think tanks, R&D centers (design-oriented accelerators), residences and studios for artists and creatives, commercial spaces dedicated to certificate products of craftsmanship and food industry, both local and foreign. For the traditional spaces of cultural production, the strong idea was to draft a program of activities (exhibitions, artistic performances, concerts, conferences, etc.) with reduced costs but an high scientific quality.

In spite of the wide availability of heritage, the report recognized how it was impossible, or at least not advisable, to host the entire spectrum of functions within the historical urban landscape. In this perspective the report proposed to enlarge the boundaries of the urban creative and cultural ecosystem to the whole region. According to the report, this regional system of production networks and chains, horizontally integrated, was the better way to manage cultural heritage and landscape in an advanced cultural district. On the contrary, the report opposed to the 'classical' district:

> These forms of reorganization are not generally guided by a planning or an institutional strategic direction. [Classical districts rather] represent spontaneous evolutions of old models in spite their growing competitive inadequacy […] a re-edition of the traditional district model (vertically integrated) applied these time to Italy's "hidden treasure", its cultural heritage. Thus, the idea of cultural district as a prolongation of the industrial district to the sector of touristic valorization of the heritage, with respect to which Italy can boast a "dominant position" […] with the development of local touristic-cultural systems. In any case, it is worth noticing how the revisitation, in the cultural sense, of the classical industrial district isn't based on a strong foundation (unless one seriously feels that the economy in cultural merchandising and artistic craftsmanship can be compared and seen equal to that of the textile and mechanic industry). In addiction, the industrial districts are autonomously created by local entrepreneurs while the cultural districts are often created by local administrations excluding external entrepreneurs. These are not trivial differences, since they explain the appalling economic of the few examples of concrete application of this "mechanistic" setting (Unioncamere and Symbola, 2012).

In 2013, the report of Unioncamere and Fondazione Symbola and the report of Federculture took over the theme of territorial networks and of governance in the creative and cultural sector, formulating various proposals. Among these, there was the creation of a Fund for Cultural Project Development, the reorganization of the competences shared between the MiBAC and other Ministries (integrate different aspects of culture and creative industry, strengthen and harmonize the role of the Regions) and the inclusion of MiBAC in the table on the Digital Agenda. Moreover, they illustrated the challenges of the 2014-2020 European programming like the access to the European Guarantee Fund and the battle of the European Council of Ministries to include the sectors of culture and creativity in the priorities of the Structural Funds. Particular attention was given to *Italia 2019*, the document that guided the selection of the European Capital of Culture 2019 among the Italian candidates[22]. Aware of the fact that the funds were lower than those for Genoa (ECoC 2004), Federculture, Unioncamere, and Fondazione Symbola presented the initiative as an occasion to re-brand Italy for the art, culture, history and, more in general, the quality of life. In particular, this choice is also coherent with the theme of Milan EXPO 2015 and the National and European programming for the period 2014-2020 (Argano, 2013; Unioncamere and Symbola, 2013).

The concentration of funds on few themes and projects is a *leitmotiv* of the government's action and an over-imposed choice due to the delay in the spending of European and National funds for 2007-2013. In the *Programma Operativo Interregionale - Attrattori culturali, naturali e turismo* (POin, Interregional Operative Program) dedicated to cultural interventions in the Southern Italy. For instance, the delay in the realization of the projects almost led to the loss of 1.5 billion euros in ERDF funds. The lack of spending of the resources coming from the Structural Funds and from the Fund for Development and Cohesion, above all in the South of Italy, required three interventions of redistribution by the central government. The first intervention coincided with the National Action Plan for Cohesion (ACP) of May 2012, wanted by the then Ministry for Development and Cohesion, Fabrizio Barca. Thanks to this plan, 330 million euros for the protection and valorization of several national cultural centers have been reassigned. The second intervention, dated June 2013[23], consisted in the

---

22 The document derives from the encounter between *Associazione Città d'Arte e Cultura* (CIDAC, Association of Art and Culture Cities), the MiBAC and the representatives of the twenty candidate cities that took place in Rome on 14 November 2012. The candidate cities were Aosta, Bergamo, Mantua, Venice, Ravenna, Urbino, Pisa, Siena, Grosseto, Perugia with the places of St. Francis of Assisi and of Umbria, L'Aquila, Lecce, Taranto, Caserta, Matera, Reggio Calabria, Palermo, Syracuse, Erice, Cagliari. On this occasion, the list of the cities that passed the pre-selection has been communicated: Cagliari, Lecce, Matera, Perugia, Ravenna, and Siena. At the end of the procedure, Matera won the competition.

23 See L. 98/2013 and L.99/2012. The former, in particular, determines the obligation for state administrations to yield to the procedures, measures, and acts relative to the activities in any way connected to the use of Structural European funds.

reduction of national cofounding of the NOP of the Convergence Objective for 1 billion euros. These resources, integrated with funding from the national ACP, have been destined to measures for youth employment and poverty. To accelerate the expense, special units[24] were instituted at the Department for development and cohesion. The third reassignment, which took place in December 2013, supported several initiatives dedicated to the development of the local economy and to projects for employment and the war against poverty (Trigilia, 2014). Considering cultural and creative economy and the related spatial policies, the most important funds are those of the Guarantee Fund for small and medium businesses and the National Plan for the City[25]. One billion euros were destined to the valorization of historical, cultural and environmental heritage in order to promote tourism for the Expo 2015. An important characteristic of all the events has been the possibility for them to be refinanced with the 2014-2020 programming.

The *Partnership Agreement for 2014-2020 Programming*, approved by the government on 22 April 2014, includes a reorganization of management of the Structural Funds and of the national ACP with the creation of the Agency for Territorial Cohesion[26]. The goal of this Agency is to rediscover the necessity of cohesion policies with respect to sector policies. In the less developed areas of the South, the 2014-2020 territorial development policies aim at the expansion of the innovative productions. The idea is to take advantage of the opportunities for certified high-quality productions and for a tourism connected to cultural and environmental heritage.

Among the recently finished projects are the Forum dedicated to the small and medium European cities in the semester of European presidency (July-December 2014) and the experimentations for territorial development of the South through the *Metro PON* dedicated to Metropolitan Cities, the Urban Agenda, and the National Strategy for Internal Areas. In addition to this, the experimental project *Cultura in movimento*[27] (Moving culture) (Argani, 2013; Kern and Montalto, 2013; Trigilia 2014). The goal was the creation of partnerships with local organizations in order to obtain a united offer by the start of the next tourist season. The cities involved (Benevento, Caserta, Catania, Cosenza, Ivrea, Lecce, Matera, Ragusa, Syracuse) took

24 For instance, task forces for Campania and Sicily have been created.

25 For interventions included in the ROPs and with the closing of the works within December 2015.

26 See L.125/2013.

27 The program has been launched on the 30th of January 2014 from the Minister for Territorial Cohesion and the Ministry of Cultural Asset and Tourism (MiBACT), born with the D.L. 136/2013 (converted to L.6/2014) that transferred the competences of the Ministry for Tourism at MiBAC.

part in the project *L'Expo 2015 nelle Regioni* (Expo 2015 of the Regions)[28], backed by the Ministry for Territorial Cohesion in 2014. The project collected ideas and projects from the Italian regions to display in at Italian Pavilion of Expo 2015 and at the following events[29]:

1. A cycle of events held in the 95 capital cities of Italian Provinces and about quality of certified products of these territories;
2. The three-days conference *ANCIperEXPO2015* at the same time of the 23rd Annual ANCI Assembly;
3. The program *15 hoursANCIperEXPO2015* composed by twenty regional events hosted by the selected cities and promoted by Anci-Expo with the involvement of national testimonials;
4. The twin towns program (from January 2014 to April 2015).

Some initiatives already organized by the Cities have been harmonized to the EXPO2015 themes and inserted in one circuit of national promotion. The Ministry for Territorial Cohesion and MiBACT have invested large resources on Expo 2015 hoping in a re-launch of the national image abroad. The choice of vertical coordination (commissioner, agencies, departments, etc.), however, limited the decisional power of the local players. Thus, they defined EXPO 2015 as a huge operation of territorial marketing by the State far from any contemporary place branding strategy[30].

The *Expo 2015 nelle Regioni* project, as other joint initiatives of MiBAC and the Ministry for Territorial Cohesion, express the desire to accelerate the spending of national and European funds in the attempt to build adequate policies for culture and creativity, for which Italy is remarkably late. It is enough to think that, in spite of a rich literature and an almost continuous statistical production on the field, *Italia Creativa* (2010) represents the only concrete project dedicated to creativity up to now. In addition to this, the economic models proposed by Santagata (2007) and Pier Luigi Sacco (2013a, 2013b) have only been applied in a few cases. This results in the cultural and creative sector being neglected from an administrational point of view still remain exclusively the responsibility of the MiBACT,

[28] The project is born after the encounter with *Le Regioni d'Italia per Expo Milano 2015* (The Italian Regions for Expo Milano 2015) which took place on 1 July 2013 in Milan and the operational encounter of 7 March 2014, always in Milan, that involved the CABINA DI REGIA EXPO 2015, the Regions, and ANCI. See http://www.padiglioneitaliaexpo2015.com/it/news_eventi/padiglione-italia-e-regioni-al-lavoro-per-expo-2015 (Official Website of the Italian Pavilion).

[29] http://www.anciperexpo.it/ (webpage of the ANCI site dedicated to EXPO 2015).

[30] The most advanced place branding strategies are based on the horizontal coordination of local players according to a precise partnership building (Anholt, 2011; Hankinson, 2010).

even though the initiatives frequently require the involvement of different Ministries, e.g. in the National Plan for the City (MIT - MiBACT) and in *Italia Creativa* (MiBACT - Ministry of Youth Affairs).

Lastly, as stated by Barbati (2013) in a comment to the Decree "Valore Cultura"[31] (Culture Value), the precariousness of the political context has slowed and limited the State's action such, has delayed the expenditure for the 2007-2013 programming, has created the continuous redistribution of the responsibilities of the various Ministries (the MiBAC that became MiBACT, the birth of the Ministry and Agency for Territorial Cohesion, etc.), has continued the long-standing dispute regarding the use of emergency decrees instead of the use of proper channels[32].

In spite these issues, the Italian context presents several examples of good practice on a regional scale which respond to the model of the advanced cultural district. The following paragraph deals with these experiences, in the conviction that these latter provide a concrete base for the constitution of creative spatial policies for cultural heritage and landscape in Italy.

3.4 The Italian context: the regional dimension

Since 2000, the Italian Regions showed an increasing interest towards the economies deriving from the valorization of cultural heritage and landscape. Despite this, at 2015 the framework of regional policies for cultural districts is still fragmented.

---

31 D.L.91 / 2013 converted with amendments into L.112 / 2013.

32 In the original text: "Nelle "disposizioni urgenti per il rilancio del cinema, delle attività musicali e dello spettacolo dal vivo", dettate dal decreto in esame, non vi è pertanto traccia alcuna delle tante soluzioni che, nel tempo, hanno accompagnato la riflessione e le proposte della dottrina e delle sedi istituzionali per la valorizzazione-promozione di queste attività culturali. […] L'insieme delle misure introdotte […] si risolve in una serie di micro-interventi, di natura essenzialmente finanziaria e, talvolta, solo fiscale destinati a favorire talune iniziative o fasi dell'organizzazione e produzione delle diverse attività. […] Si tratta, in realtà, di "piccole" misure che valgono soprattutto a rimuovere taluni adempimenti che le modificate condizioni dei contesti, anche normativi, inducono a ritenere non più attuali e perciò esigibili. Piccole misure, quando poste a confronto con la complessità e la complicazione delle procedure ancora previste per molte fasi della produzione e organizzazione di tali attività. […] Nel decreto "Valore Cultura", "qualche cosa" c'è ma questo "qualche cosa" è espressione delle "piccole possibilità" di una politica stretta fra l'urgenza di intervenire, anche per segnalare il proprio ingresso in scena, e l'incertezza in merito agli spazi temporali e materiali consentiti alla propria azione. Quell'incertezza che, pur non essendone unica causa, concorre a generare la debolezza che non consente pensieri e interventi di sistema specie per i settori, com'è quello della cultura, che non si riescano o non si vogliano attrarre nell'ambito delle politiche pubbliche generali" (Barbati, 2013).

The districts set up thanks to the regional cultural policies are only twenty: the first place goes to Lombardy with 6 districts, followed by Marche with 3 and by Veneto with 2. The Regions that took part in the call on technological districts by the Ministry of Education, Research and University - MIUR (National program of Research 2005-2007) host a technological district each.

The distribution of the districts in relation to the creation mechanisms is homogeneous with 4 districts born thanks to the MIUR call, 4 thanks to the regional laws on industrial districts and 3 thanks to the regional laws on cultural districts. The legislative intervention by Regions does not represent a decisive element, since in the regions without a specific discipline 10 cultural districts have been created (6 in Lombardy, 2 in Molise, 1 in Piedmont, 1 in Trentino). The success of regional policies for districts depends on the presence of a three-years work program with a clear strategy behind, as for Friuli Venezia Giulia, Sardinia, and Marche regions. In fact, cultural districts have political implications which, in the case of cultural heritage, seem to be perceived as an "uncomfortable fact" by scholars and administrators, above all in the assignment of the district feasibility study to external consultants (Ponzini *et al.*, 2014).

The national system of advanced cultural districts is composed by technological districts, meta-districts for restoration and preservation, districts for cultural tourism, districts for wine and food, traditional cultural and museum districts. These latter are the simplest forms of cultural networks. They can be the result of the old regional policies for culture or, instead, the outcome of some recent projects on cultural/tourist districts that were not able to evolve remaining at a rudimentary stage, as for the *Distretto Culturale del Sud-Est* (Cultural District of the South-East) in Sicily. Besides the aspects of promotion and territorial marketing, this district did not sponsor new forms of cooperation among the players of the cultural heritage chain in the Late Baroque Towns of the Val di Noto (Carta, 2003).

At 2015, the districts that can be defined as "advanced" as "systems of systems" are the DiT-BeCs and *Puglia Creativa* districts in Tuscany and Apulia, and the cultural districts funded by Fondazione Cariplo in Lumbardy (even if these latters differ for a mono-chain approach and for multiple integration that mainly concern the public administration) (Montella, 2015; Nuccio and Ponzini, 2016; Ponzini, 2016; Usai, 2015).

Considering the policies for the historical and existing city, the places of culture (i.e. museums, research centers, archives, etc.) represent the focus of both district policies and landscape planning[33]. Despite this, the

[33] The reference here is to landscape planning as defined by the European Convention of Landscape: "*'Landscape planning' means strong forward-looking action to enhance, restore or create landscapes*" (Council of Europe, 2000, art.1 letter e).

urban planning is taken into consideration only in the redefinition of the destinations of the heritage assets and the landscapes to be recuperated. This is the case of the Regge dei Gonzaga, the Barco Ducale of Urbania and the Polironiano Complex in San Benedetto Po in the cultural districts of Lumbardy by Fondazione Cariplo (Barbetta *et al.*, 2013; Camerlengo, 2013; Papi, 2014). Nonetheless, only the Regional Landscape Plan of Campania Region took care of the construction of cultural districts (even though according to a tourist approach).

Keeping in mind the experiences analyzed, within the advanced cultural district it is possible to individuate six different models of "creative network": the cultural network, the local cultural system, the technologic cluster, the cultural district, the advanced cultural district, the meta-district (Alberti and Giusti, 2009; Seddio, 2013). They coincide perfectly with those present in the advanced industrial district, with the only exception of *supra*-local clusters that, in the cultural field, do not manage to establish autonomously. The cause is a too institutionalized and institutional vision of the cultural district which links the creation of *supra*-local clusters to the internationalization policies adopted by public administrations.

In addition to this, the cultural districts often weave with other forms of advanced district, as in the case of the districts for food and wine (Piedmont), tourist districts (Campania, Molise, Sicily, and Trentino) or technological districts for restoration and preservation (Tuscany). This allows to identify the advanced cultural district as a particular form of the creative industrial district, characterized by:

- an organization with variable forms of management;
- systemic interdependencies;
- choices of vertical, horizontal and cross-cutting integration between the cultural heritage chain and the other industrial chains.

The advanced cultural district creates new occasions of training, learning, innovation and creativity for the local productive systems that, nonetheless, rarely give life to *supra*-local clusters (with the only exception of the meta-districts) (Montella, 2015; Nuccio and Ponzini, 2016; Ponzini, 2016; Usai, 2015).

Urban policies and, more generally, the tools of urban planning have a marginal role in district policies which is limited to the definition of new uses for the heritage assets and the landscapes to be recuperated. It is a unexpected data considering that, with the DLgs.42/2004, the regions had to face in urban planning the same issue posed by district policies, i.e. revising and update their policy and administrative system to the new principles and rules on cultural heritage and landscape. Some regions have modified the existing plans inserting specific rules for landscape. Other regions have adopted new plans, called Regional Landscape Plans.

The outcomes have been different according to the ability of regional administrations in handling the changes of legislation, the relations with local organizations and, above all, the dialogue with the MiBACT's offices. In 2014, ten years after the adoption of the DLgs.42/2004, Tuscany and Puglia are the unique regions which have concluded the co-planning with MiBACT and that have adopted their Regional Landscape Plans. The same regions that emerged in district policies as examples of good practice, confirming that the success of every initiative does not depend on the measures implemented but, on the contrary, on the adoption of a clear and transparent strategy in the sector.

3.5 Conclusions

The experiences of cities and regions on a global scale offer different interpretations of cultural and creative clusters/networks.

In countries where institutions and public organizations show a dual attitude towards cultural and creative industries (considered, a new element of welfare and, at the same time, an economic sector to develop), cultural and creative networks/clusters are analyzed according to their mission. This approach put in contrast the clusters and networks which support social inclusion with those supporting entrepreneurship (social purpose vs. economic purpose). It investigates the conflicts and the frictions between the respective groups of interest (citizens and entrepreneurs) and how they influence the interventions of urban renewal in a direction or another one. This approach is common in the Anglo-Saxon world (United Kingdom, Australia, United States) and in China, however the spatial effects are more evident in Canada, in the United States and in Asia as well, for emulation (Stevenson, 2014).

In Canada, France, Belgium and, for some aspects, even in Australia, cultural and creative networks/clusters are commonly investigated on the base of the *top-down* or *bottom-up* nature of their projects while, in post-colonial or post-communist countries (Australia, South Korea, Hong Kong, Singapore, and Baltic Republics[34]), cultural and creative networks/clusters are often studied considering the adherence of interventions to the local communities' culture (Mercer, 2008; Rozentale and Lavagna, 2015). In this case, the studies on cultural and creative industries focus on: the ideologies and the narratives that permeate the policies destined to/produced

[34] On this topic, see Mercer, (2008) and also Rozentale, I. and Lavagna, M. (2015). The "universal" characteristics of creative industries revisited: The case of Riga, in *City, Culture and Society*, 5(2014), pp. 55-64.

by cultural and creative networks/clusters; their relations with the pre-existing cultural and creative systems; their social impacts for local communities (Ferilli *et al.*, 2016). Usually these studies operate the following distinction: 1) initiatives that do not take local identity into consideration and obstacle the construction of spaces of dialogue, following the project deliberated by the commitment (China, South Korea); 2) initiatives that take into account local identity including it in the project *vision* from the beginning through the concentration of goals and interventions (Europe, Canada, Australia) (Girard *et al.*, 2016).

In all countries already mentioned, however, it is possible to perceive a shift from an organizational model pre-eminently urban (cultural precincts, concerning the neighborhood or the quarter), towards a regional one which involves different types of actors and spaces joint together in a "network of networks". This new model is increasingly analyzed and governed through the archetype of the "advanced industrial district" in its different forms (cultural industry district, advanced cultural district, creative district). It is a productive and relational model born from the Italian experience of small and medium enterprises and it is characterized by openness and inclusivity. Despite this, the advanced industrial district presents some important ethical issues as the previous organizational models. For example, the social and spatial conflicts connected to the application of "global" production schemes at the regional and urban scale, as proved by the extensive design of cultural districts in emerging Asian metropolis (Comunian *et al.*, 2014; Ferilli *et al.*, 2016; Girard *et al.*, 2016; Hutton, 2016, pp. 194–134). Equally meaningful is the admonition coming from Italy where the absence of a clear and transparent policy frame for culture and creativity is encouraging the insurgence of "advanced cultural districts" on the whole national territory reducing the effectiveness of spatial and cultural policies at urban level (Montella, 2015; Nuccio and Ponzini, 2016; Ponzini, 2016; Usai, 2016).

For this reason, the following chapters are dedicated to the advanced cultural districts, their history and theoretic foundation but, above all, to their potentials in the definition of a global taxonomy for cultural and creative networks and clusters and in urban planning.

Regional Planning

The Importance of the Advanced Cultural District

4.1 The origins of the advanced cultural district: the industrial district

The industrial district represents a complex object of analysis, which can be read under different but complementary perspectives. According to Antoldi (2006) and Cerrato (2008), they are:

- the *economical perspective*, which looks at the districts as a peculiar form of industrial organization, alternative to the fordist model of large companies;
- the *corporate perspective*, more directly connected to the strategic and organizational profiles of district businesses;
- the *socio-cultural perspective*, which focuses on the social capital of districts as a component of their production structure;
- the *cognitive perspective*, which emphasizes the learning mechanisms and knowledge transfer among the actors of the district.

These perspective are taken as a reference in the following paragraphs to illustrate the history and theoretic foundation of the advanced cultural districts.

4.1.1 The classical district: a new economic-commercial approach in industrial organization (1920-1970)

In the Postwar period the neoclassic economy provided the theoretical basis for a model of growth connected to the tangible aspects of the production (invested capital, workforce, technical innovation – tools, means, and procedures). Social capital was considered an irrelevant exogenous factor or, in some cases, an obstacle to modernization[1] (Giordani and

[1] This is the case of economic programming in the South of Italy around the 1950s and 1960s to close the gap between North and South, considered the break to national development. This double reading of the country from a territorial point of view (industrial triangle *vs.* the rest of Italy), sectorial (industry *vs.* agriculture), and dimensional (large industries *vs.* small and medium businesses) created state policies to support urbanization and industrialization, seen as unique and self-propulsive mechanisms for

Zamparelli, 2007). In spite of that, since the 1920s the vertical integration of the fordist model was questioned by the studies of Alfred Marshall (1879, 1919, 1920) which introduced the district model in the studies on industrial organization and technological innovation. Basically, Marshall observed how the advantages of production on a large scale could be obtained also grouping in a narrow geographical area a system of small producers, each responsible of one of the phases of the productive cycle (invention, production, distribution). The topic of industrial district by Alfred Marshall placed among the traditional concepts of "company" and "sector". It is constituted by a set of businesses of the same sector, localized in a circumscribed area, among which there is a local strengthening of collaboration, but also of competition (Dossena and Cioccarelli, 2013). In the district model, business relationships became a relevant factor as a peculiar expression of a historically and geographically determinate "culture of production". Beyond the traditional vertical relations (hierarchy among the phases of the same production), there were horizontal relations (more companies carried out similar activities in the same productive process) and diagonal relations (more companies produced ancillary services to the district) (Bellandi, 1987; Bondonio and Debernardi, 2006; Palmi, 2013).

4.1.2 The growth of the socio-technical dimension within the economic-commercial approach (1970-1990)

In the 1970s and in the 1980s the social dimension of the industrial district acquired relevance in relation to the studies on local development[2] (Garofoli, 1983). In Italy, Bagnasco (1977), Becattini (1975), Brusco (1980), Anastasia and Rullani (1981) individuated multiple configurations for the industrial district. Each of them respondent to the territorial capital, i.e. the set of endogenous factors relevant for the production (goods, resources, values, habits) and their combination on a local level (links among economy, society and institutions) (Camagni, 2009). Between the *Milan-Turin-Genoa industrial triangle* and the backwardness of the Southern Italy, there was the Third Italy of industrial districts with the blooming of small and medium businesses in

---

the modernization of the South of Italy. In this context, culture and "community" factors have been taken into consideration only by those who were engaged in the study of the depressed areas of the country, while the ruling class (economic and political) considered them as an obstacle to socio-economic change and national economic policies (Trigilia, 1998).

2 Thanks to these studies it was possible to identify the community factors influencing the economic growth (economic role of the family, of free-lancers, of small businesses, of non-economic exchanges based on reciprocal favors, etc.)

the North-East, in Emilia-Romagna, in Tuscany and in Marche regions (Bagnasco, 1977).

The dimension of the phenomenon in terms of employment and income attracted the attention of economists, sociologists, planners, and administrators who investigated the reason of success of these unattractive geographical areas (hard accessibility, no mature industrial productions, prevalence of family businesses) (Becattini, 2002).

According to Rullani (2002) the crisis of Fordism transformed some factors of disadvantage in factors of growth, accelerating the passage from a model of development base on accumulation to one of expansion (Palmi, 2013, p. 26). According to Becattini (1989) and to the following studies[3], the Italian districts" are born thanks to the convergence of some collective practices rooted in the local communities, which allowed a balance between individual and collective interests (Becattini, 2002; Bellandi and Sforzi, 2001; Dei Ottati, 1995; Sugden, 1989). Thus, a population of enterprises and the geographical proximity are a necessary but not sufficient condition for the birth of a district. Instead, the key element is the presence of a community founded on a homogeneous system of values and practices (sense of belonging, reciprocal trust, ethic of work, knowledge and behaviors of the *homo distrectualis*) (Sacco and Pedrini, 2003).

Becattini (1989) took over from Alfred Marshall's definition of district and industrial atmosphere stressing the socio-technical dimension of the production system[4]. Becattini's industrial district, shaped as a unique and not reproducible functional organization, is based on these features (Capello, 2004; Hinna and Seddio, 2013):

1. the complexity of social and economic relationships;
2. the territory as a *medium* between the enterprises and the local community;
3. the social constraints of the local community (credibility and reputation),

---

[3] See Bellandi (1995), Cozzi (2000), Folloni and Gorla (2000) and Lazzeretti (2008, 2011).

[4] This tendency can also be found in spatial planning. In fact, it is possible to register a shift from rationalist plans to planning tools made with the involvement of local people. In particular, the reference is to urban plans inspired by the studies on urban citizenship and on the ideal city (the *good city*) by: the British *Garden City and New Towns Movement* linked to the work of Ebezener Howard, Patrick Geddes, Patrick Abercrombie and Lewis Mumford; by the *City Beautiful Movement* of the School of Chicago (Frederick Law Olmsted and Daniel Durham); by the studies of Jane Jacobs (1961, 1969) and Sharon Zukin (1995, 2008, 2010). On the topic see Carta (1996), Di Biagi (2009), O'Connor, and Shaw (2014).

4. the social proximity, i.e. the closeness of businesses to institutions, research centers and Universities.

For instance, the empiric experience showed how the enterprise population of a district can vary for dimension (concentration of large businesses, medium, and small) and for attitude (predatory or colonization attitude, peer relations of collaboration and exchange, leadership positions) (Palmi, 2013). On this basis, Rullani (2004) defined the industrial district as an "extraordinary form of involuntary cooperation" while Hinna and Seddio (2013) discussed it in relation to the progressive revaluation of territorial capital.

4.1.3 Industrial districts in the global market: a socio-cultural approach in industrial organization (1995-2005)

In the 1990s, the Italian districts protected the industrial sector from different recessive trends but since 2000 they have shown signs of inadequacy with respect to the international economic context. The globalization, in fact, has determined an increase in economic competition and in market turbulence in all fields. This is due to macro and micro factors which literature has not been able to identify yet[5]. Among the macro-factors there is surely the competitive prices of manufactures in emerging countries. This moved the attention of the Western world on the phases before and after the production. Knowledge, creativity, and innovation became the determinant factors of the new growth. If "do it" and the "do it better" are no longer the prerogative of a specific geographic area, "do it earlier and differently" becomes the *leitmotiv* of the new industry (Hinna and Seddio, 2013; Palmi, 2013). Mobility, information and communication technologies (ICT) had an equally fundamental role. The new technologies, in fact, made the relation within and without the district more rapid and efficient reducing the economic advantage offered by spatial proximity and increasing the advantage coming from the cognitive, organizational, social and institutional closeness (Table 11) (Bonacci, 2007; Boschma, 2005).

[5] In Italy the involution of the district model is due to the deep changes that took place in the previous twenty years: the reform of the institutional system in favor of local autonomies (L.142/1990) and the reform on services in museum and public infrastructures for culture (L. 142 e 241 del 1990, L. 4/1993 - *Ronchey Law on cultural and museum services*).

Table 11. The relationships between enterprises in the industrial district

Theoretic Approach	Coordination mechanism	Main focus	Function of the enterprise
Classic and neoclassic economy (1970s)	Market	*Allocation*	Maximization of performance
Development of mass production (1980s)	Hierarchy	*Coordination*	Maximization of performance
Reticular capitalism (1990s)	Plurality of organizational relations	*Relationship*	Maximization of scientific and technological knowledge
Economy of access (2000)	Polymorphism of organizations	*Relationship*	Rational effectiveness

Source: Bonacci (2007)

This operational context, more dynamic and fickle, accentuated the crisis in manufacture districts with a strong focus in subcontractor. The main element of weakness is here the scarce inclination to innovation deriving from (Varaldo, 2006):

1. Defensive behaviors: high grade of self-sufficiency and self-referentiality in term of competences and skills, penalized external collaboration, limited renovation and enrichment of social and cultural capital (lock-in effect) (Boschma, 2005; Palmi, 2013);
2. Localism: an excessive emphasis on the inner qualities of the local environment (business and district distinctiveness is linked to the place of origin and not to the work of the companies within and without it – let us think, for instance, about the crisis of the *Made in Italy* label) (Hinna and Seddio, 2013; Perrone, 2004);
3. Atomization of the productive tissue: the presence of few big companies in a production systems made of small firms which inhibit the accumulation of knowledge and the dimensional growth (Becattini, 1979; Salvemini, 2008);
4. Scarce consideration of the cultural and creative dimension of the productive process, also from a financial point of view, in terms of investments in research and development, communication and protection of intellectual property (Palmi, 2013).

The last point is particularly important because it explains the success of the few Italian districts that prosper.

They are those who gave value to the protection and commercialization of inventions and understood the importance of intangible factors, such as information and telecommunication (Palmi, 2013). This happened after the growth of national and international companies that chose to settle in existing industrial districts and, at the same time, the expansion of the already existing companies through the introduction of new knowledge.

This phenomenon, named cross-fertilization or location paradox, deeply modified Marshall and Becattini's industrial district determining a change in:

1. the relationship between large companies and small-medium enterprises within the district. This happens due to:
   a) the experimentation of different trajectories of development by some companies which assumed the role of leaders and developed plans for expansion independently from their district;
   b) the choice of other companies to place themselves under the protection of the first ones, taking care of manufacturing, although less profitable (Hinna and Seddio, 2013, Palmi, 2013);
2. the relationships between the companies of the district and the rest of the market with:
   a) the diffusion of networks in enterprises, meaning that the groups of companies operating in connected sectors became allies and put activities and resources together to strengthen their production, while not merging the management of the companies;
   b) the diffusion of the network-enterprises, meaning a form of organization where a main company assigns the various phases of production to other subcontractors. Although this "main company" is the one to assign the work, it does not have leadership over the others. Defined in Latin as "*primus inter pares*" (Dossena and Cioccarelli, 2013; Soda, 1998).
   c) the concentration of district production in the hands of a small number of leader enterprises, which ended up controlling all the production.

Economic literature analyzes the phenomenon of clustering under various aspects; it takes into consideration the factors of decline and transformation, the internationalization of the technological clusters, the entrance of new actors (enterprises, investors and knowledge brokers), the adoption of new technologies in order to compete with the rival enterprises, the presence of combined economic and social inter-organizational bonds which influence the performance of the clusters (Boari, 2010).

This last factor, called embeddedness, is particularly important because it shows the role assumed by the relational dimension, by the governance[6], and by the territory in which the industrial district operates (Dematteis and Lanza, 1999, p. 37). Analyzing the relationships between the district model and culture in the tourist field, Valentino (2003) found an important "lever" for local development in the valorization of the cultural heritage. He also defined the historical-artistic and environmental assets as "nodes" for the constitution of integrated tourist local systems. The model elaborated by Valentino (2003) provided interesting points for a new interpretation of the industrial district as a network[7] (Hinna and Seddio, 2013, p. 37; Williamson, 1975, 1991).

In the figurative use of the model, the "levers" for the development became the ability of the district to self-organize as a local territorial system[8] basing it on the socio-economical-cultural context (Debernardi, 2005). With Valentino's studies (2003) the economic and corporate vision of the neoclassical theory left space for a new approach focused on the socio-cultural dimension of the district (Palmi, 2013, p. 59).

The territory was analyzed in literature on endogenous development under various aspects: first as an asset of knowledge and then, after Dematteis and Lanza's studies on the *milieu innovateaur* (1999), in terms of social, cultural, relational and territorial capital, offering a solid theoretic base for the construction of an "advanced" or "creative" industrial district (Camagni, 1999, 2004, 2009; Camagni e Capello, 2002; Foray, 2000; Putnam, 1993; Storper, 2003).

---

6 The concept of *governance* is here used in relation to a system of cooperation that involves actively public institutions, organized groups, and local players to reach common objectives shared in the analysis (Seddio, 2004, p. 123).

7 A qualifying network above all for medium, small, and very small companies that, alone, would not be able to compete on a global scale (Palmi, 2013).

8 The local territorial systems are territorial socio-cultural entities made by the following elements: the local network (a local aggregation of subjects that acts as a collective player); the local *milieu* (the set of physical and socio-cultural conditions that sediment in time thanks to shared local projects); the ecosystem (natural environment of reference). The studies have recognized to territorial local systems an ability of self-regulation and balance between individual and collective interests coming from (1) interaction of local milieu and ecosystem, consisting in translating the potentialities of the milieu in communicable and exchangeable values, through processes of symbolic and material transformation of the environment; (2) the interactive relation of the local network with *supra*-local networks, that allows to "export" the values produced by interaction between local network and milieu (Dematteis, 2001; Debernardi, 2005).

4.1.4 Advanced industrial districts: a cognitive approach in industrial organization and technological innovation (2005-2014)

In recent years, the industrial district has been reshaped for the cultural production according to the studies on cultural districts and creativity (Palmi, 2013, p. 67). In spite of this, it is still difficult to talk about a perfect coincidence between productive and organizational models of the cultural district and those of the "classical district", because cultural and creative industries mainly deal with immaterial goods and services, and because not all the territories have a concentration of cultural and natural assets that can guarantee the levels of production, employment and income of traditional industries[9] (Palmi, 2013; Salvemini, 2008, p. 8).

Culture and art, considered in their broadest meaning[10], represent an essential input for the development of advanced industrial districts, i.e. "creative districts". However, there are other factors that must be considered (Palmi, 2013; Sacco and Ferilli, 2006):

1. the quality of life determined by the presence of practitioners and talents of the creative class ("*the world is spiky*") (Florida, 2002);
2. the inclination towards innovation in the transition towards a post-industrial economy (Porter, 1990, 2003);
3. the cognitive capacity and the motivational re-orientation, i.e. the expansion of the individual experiences (functioning) and alternatives (capacitation) (Sen, 1994, 2002).

Creative districts are territorial local systems which offer a competitive advantage thanks to the creation and diffusion of new knowledge[11]. The organizational studies on innovation have outlined the learning processes which characterize the creative districts and they are (Palmi, 2013):

– *exploration*: learning processes aimed at identifying new development models;

---

9 This is possible in cities like Rome, Florence or Venice. The massive cultural tourism in Venice, however, teaches that even the "economy of valorization" can be counterproductive if not accompanied by a clear urban strategy aimed at residents, investors and visitors.

10 According to the definition adopted in cultural programming after the 1990s, culture is defined by taking into account all the tangible resources (sites, buildings, artworks and artifacts) and intangible resources (attitudes, practices and beliefs) necessary for the functioning of the district as a society (Throsby, 2001).

11 Even if learning is now linked to the simultaneous action of endogenous and exogenous factors, related to the phenomenon of glocalization (Bauman, 1998; Wellman and Hampton, 1990, 2002; Robertson, 1992)

- *exploitation*: learning processes aimed at strengthening knowledge and skills in possession of companies and at incorporating them within the production chain;
- *absorption*: learning processes deriving from the identification, acquisition and implementation of creative knowledge owned by external actors[12].

The district is places at the center of wide networks. This requires the construction of *supra*-local clusters respect to the existing ones (groups of enterprises, networks of enterprises, network enterprises). The meta-districts and the networks of districts, born thanks to ICT and to laws on the matter[13], have modified the governance of industrial and cultural districts. The meta-districts are networks of enterprises operating in different places which are complementary in terms of productions and skills[14], while the networks of districts connect districts that are complementary only under the technological point of view[15] (Palmi, 2013).

The construction of a multi-level network is a complex path that requires a clear definition of the roles assigned to the different territories and a form of government, fundamental to provide strategic unity and an organizational coordination of the network (Palmi, 2013). For this reason, district literature tends to focus inexorably on the phenomenon of grouping trying to identify, on the one hand, the natural and voluntary mechanisms that induce enterprises to aggregation, on the other, the institutional (artificial) mechanisms able to influence and encourage local and *supra*-local clusters.

In the first case, the creative district is analyzed as an economic cluster, close to the classic industrial district. In the second, it is analyzed as a specific action of policy that can be planned and designed (Amari, 2006; Hinna and Seddio, 2013; Palmi, 2013; Ponzini *et al.*, 2014; Santagata, 2002). After these considerations, the advanced industrial district or "creative district" can be defined as an organizational form with (Francesconi and Cioccarelli, 2013, p. 16):

- variable relational networks;
- systematic interdependencies;

12 From this process derives the absorptive capacity of a company, absolute and relative. The former refers to learning coming from confrontation with the general context, the latter from the confrontation with a rival organization. Cfr.1.6.

13 See L.R. 1/2000 of the Lombardy Region and the L.R. 8/2003 of the Veneto Region.

14 They are responsible of a particular phase of the production or they have skills not present in the other companies of the network.

15 For Italy see the L.33/2009 that introduced the network contracts (*contratti di rete*).

- vertical, horizontal and cross-cutting forms of integration which link the cultural heritage chain to the other production chains;
- connections, synergies and complementarities respect to the territory of belonging.

Despite this, the creative district is often considered as a conceptual framework for the analysis of local territorial systems rather than an industrial organization model. This is due to its cognitive approach that allows to generalize on the existing systems of relation (clusters, networks of enterprises, network enterprises, meta-districts, networks of districts), overcoming the limits of case-studies research. A new theoretic model connected to the studies on innovation that can be applied to different types of district, even the ones linked to heavy industry, chemical industry and energetic industry (see Figure 7 and Table 12).

Figure 7. *The creative district: a theoretical model*

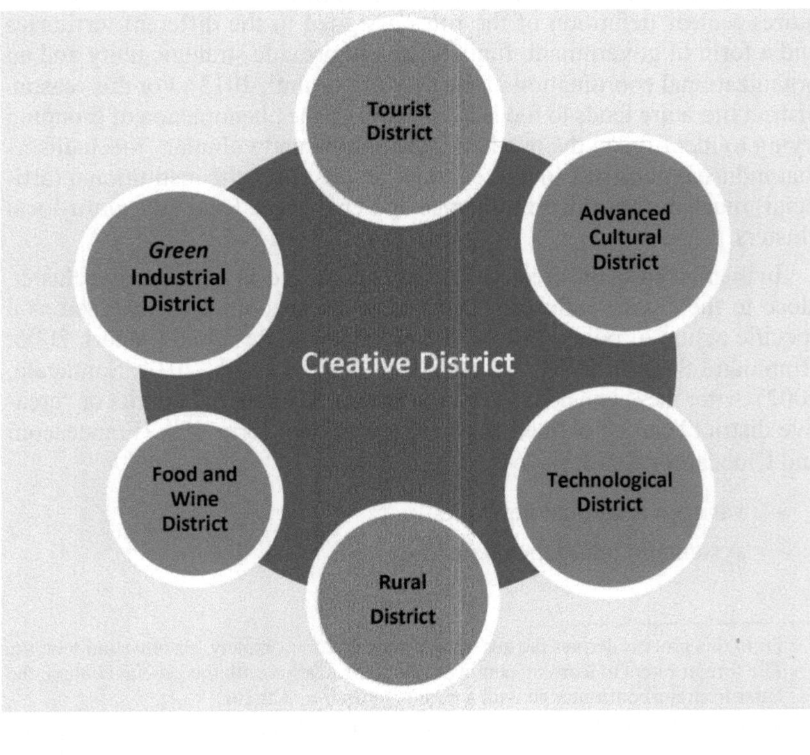

Source: elaboration of the author

Table 12. The new forms of district

Type	Definition	References
Rural districts	Local production systems characterized by a strong historical and territorial identity. They originate from integrations between agricultural productions and other economic activities coherent with the local tradition and inherent to the nature of the territory (L.D. 228/2001).	Cerrato (2008) Corsaro and Triola (2006, pp. 103-105)
Food and wine districts	Interregional production systems, characterized by networked and interdependent companies working in the food and wine sector, particularly certified and typical products (L.D. 228/2001).	Cerrato (2008*)* Corsaro and Triola (2006, pp. 103-105)
Tourist districts	Local tourist systems belonging to different regions and characterized by an integrated offer including cultural and natural heritage and tourist attractions (L.135/2001).	Amari (2006) Mariotti (2010) Santagata (2002)
Advanced cultural districts	Districts organized by local entrepreneurs and characterized by forms of protection for the cultural and creative products (intellectual property rights, commercialization, etc.), recovery and valorization of the cultural heritage and the production and circulation of innovative ideas.	Sacco and Pedrini (2003)
Technology districts	Industrial systems characterized by the presence of Universities responsible for profitable inventions in a scientific field and a network of small and medium enterprises able to develop these technological innovations.	Bossi *et al.* (2007) Mele *et al.*(2008)
"Green" industrial districts	System of "eco-industrial symbiosis" where companies and local communities cooperate to improve the economic and environmental performances of the territory of reference. They are based on the efficient use of local resources and the increment of exchanges of waste material among enterprises to reduce the environmental impact and the energy consumption by the industrial processes.	Cariani (2010) Cutaia and Morabito (2012) Fondacci (2013)

Source: elaboration of the author

4.2 From the industrial district to the advanced cultural district: the Italian contribution to the international debate

The concept of cultural district spread in late 1980s when the context was particularly favorable to an economy-driven management in the cultural field. The cuts to public budget, in fact, favored the affirmation of public-private partnerships in urban policies, of the planning for the arts and the ecosystemic theory in the studies on technological innovation.

The United Kingdom was a pioneer of the economy-driven management for culture testing its effectiveness on a large scale thanks to the studies on creative industries of the *Department Culture Media and Sport* (DCMS), to the works of Landry and Bianchini (1995) and to the dispossession of brownfield in several English cities. Neighborhoods with a high concentration of activities and places for art and entertainment, labeled as "cultural districts", spread all over the UK, thus the *Greater London Council*'s cultural strategy was recognized as the symbolic birth of the *cultural district* (Sacco and Pedrini, 2003; Sacco and Ferilli, 2006).

In the Anglo-Saxon model, however, the term "district" is used to indicate a neighborhood or an urban area, not "the industrial district" as intended by Alfred Marshall[16]. Thus, all aspects connected to the intangible heritage of knowledge and values characterizing the local productive system were not adequately taken into account. Above all, the role of the institutional networks in stimulating the grouping of local entrepreneurs though policy-making[17]. This aspect is particularly relevant for Italy where the cultural districts were developed after the application of Becattini's industrial model to the cultural heritage management. The industrial district was adapted to the cultural production through a re-vision of the inner relational networks. This explains the profound differences between the two districts (Table 13). Firstly, in the industrial district the production and consumption phases are spatially separated while in the cultural and creative industry they often coincide in space and time[18]. Secondly, the industrial district is usually the result of an "extraordinary form of involuntary cooperation" (Rullani, 2004) while the cultural district is often the output of a public policy (Preite, 1998, p. 26)[19].

---

16 The cultural district kept from the classical industrial district only the spatial concentration and the geographic proximity trying to obtain the same competitive advantages.

17 An aspect which clearly emerged in the *London Docklands* strategy in relation to the coordination activity of the *Greater London Council*.

18 If the cultural goods which can be reproduced are excluded (e.g. a record reproducing the sounds of a live concert).

19 Culture, when conceived as a product of cultural and creative industry, remains a good of public use. This makes possible that public institutions accomplish a fundamental

Table 13. Industrial and cultural district compared

	Industrial district	**Cultural district**
Purpose of the district	Valorization of a production chain	Heritage valorization within the cultural heritage chain
Importance of the socio-economic-cultural context	High	High
Birth of the district	Spontaneous or "natural"	Institutional or "artificial"
Involvement of public-private players	Mostly private players with the support of public institutions	Mostly public players with the eventual involvement of private players
Distinction between production and consumption	Present	Absent
Orientation toward innovation	Tends to be high	Tends to be low

Source: Dossena and Cioccarelli (2013)

Valentino (2001, 2003) defines the "planned" cultural district as:

> a geographically limited system of relationships which integrates the process of valorization of cultural heritage, both tangible and intangible, with the infrastructures and productive sectors necessary to it[20] (Valentino, 2001, p. 3).

The relational networks of the cultural district responds to a plurality of objectives: making more effective and efficient the cultural production; consolidating the production chain for the birth of new enterprises; widening the cultural offer; self-financing of the sector, promoting the territory of belonging. The policies of reference for the cultural district are therefore policies on accessibility, sustainability, organization and innovation (Amari, 2006).

The probability that a local network evolves in a cultural district increases when:

- culture is considered as a necessary local development factor and not as a service: this means to adopt policies for local development

role in addressing the designing of alternative models of culture-driven socio-economic development, as in the case of the Italian cultural district (Francesconi and Cioccarelli, 2013).

20 In the original text: "un sistema territorialmente delimitato di relazioni che integra il processo di valorizzazione delle dotazioni culturali, sia materiali che immateriali, con le infrastrutture e con gli altri settori produttivi che a quel processo sono connessi" (Valentino, 2001, p. 3).

centered on the heritage's fruition (empowerment of the infrastructures, of the hospitality system, of professional training, of the touristic field, of the handicraft and food industry);
- cultural heritage is considered an instrument to answer new social needs in a perspective of qualitative and quantitative development: this means to adopt policies for an adequate valorization of the remarkable cultural and natural assets;

To verify the presence of these pre-conditions in a territory, Carta (2004) proposes the following panel of indicators:

- *cultural centrality:* concentration, distribution and specificity of cultural and natural heritage assets;
- *flexibility*: offer of services according to the demand;
- *closeness and accessibility*: endowment and efficiency of the hard and soft infrastructures;
- *economic volume of the cultural heritage chain*: preservation, valorization, promotion, education and scientific research;
- *cooperation*: measures and agreements between stakeholders and social forces to have access to negotiated programming;
- *participation of the community*: initiatives for the participation and the involvement of citizens.

To summarize, the Italian cultural district is different from the Anglo-Saxon one as it considers culture both as a productive sector (cultural and creative industries and their production chains) and as a cross-cutting factor able to influence local development. This dichotomy is due, to a large extent, to Santagata's scientific contribution (2002, 2007, 2009). The author who defined for the first time the phases of the cultural heritage chain proposing a new taxonomy for the cultural districts. A classification that was able to coordinate the Italian model with the Anglo-Saxon one, obtaining a huge response on an international level[21] (Table 14).

---

21 According to Santagata's analysis, the Anglo-Saxon district model is characterized by the focus on production (cultural and creative industries) and on agglomeration processes, natural or generated through policy actions. The best representative types of districts are the cultural district and the metropolitan one. On the contrary, in the Italian case the district model is characterized by the focus on cultural heritage and its chain (preservation, valorization, fruition, management, and communication) and on the cognitive function of the territory, in agreement with the socio-technical dimension of Becattini's model. In any case, an institutional intervention, more or less marked, is expected. Therefore, the most representative type of district are the institutional cultural district and the museum one.

Table 14. Classification of cultural districts: economic-institutional features

Typologies	**Characteristic**				
	Model	**Role of public administrations**	**Sectors involved**	**Positive externalities**	**Protection of intellectual property**
Industrial Cultural District	Natural	Definition of development trajectories (no explicit public *policy*)	Design, audiovisual, cinema, fashion	Related to production	Patents, industrial secrets, tacit knowledge, trademarks
Institutional Cultural District	Semi-planned	Attribution of brands and property rights	Culture of "savoir vivre", fairs and exhibitions on agrifood industry and certified products	Related to production and consumption	Territorial brands, Designations of origin
Museum Cultural District	Planned	Cultural policies	Network of Museums	Related to consumption and networking	Copyright, trademarks (logos and store signs)
Metropolitan Cultural District	Planned	Urban policies	Theatres, cinemas, art galleries	Related to agglomeration	Copyright, authorship rights

Source: elaboration of the author on Santagata (2002) and Hinna and Seddio (2013)

Santagata's theses was critically taken over by Pier Luigi Sacco, who recognized the existence of two research paths, one linked to production and another to policy making. The author used the concept of "culture as a process" by Throsby (2001) to mediate them and create a common background for research, i.e. the advanced cultural district. This latter was defined as the result of a complex negotiation among the players of the local development system and the top-down and bottom-up forces exercised by them (Sacco and Pedrini, 2003, p. 33). A form of self-organization which answers creatively to the historic and geographical limits of the international and local context in order to attract the 'creatives', to produce innovations and re-motivate the individuals (Palmi, 2013; Sacco and Blessi, 2006; Sacco *et al.*, 2006).

In the advanced cultural district, as in all the other forms of creative district, culture is a central component of the urban regeneration strategies. In the identification "from the top" of the cultural and natural assets to be recovered or reorganized as spaces of cultural production (containers); in the management of the recovery process on behalf of institutions or private developers through participation strategies aimed at the creation of new practices, traditions and values (contents) around the properties recovered[22].

The strategies of cultural district are always focused on cultural heritage, but they are mainly centered on the quality of the cultural offer, the local governance, the production of new knowledge, the development of local firms, the attraction of talents and creative industries from outside, the participation and training of the local community, the handling of marginalization and other social problems (Sacco and Ferilli, 2006).

The relational system of the advanced cultural district is particularly interesting from a scientific point of view due to the high complexity[23].

Adopting a cognitive perspective, it can be defined as "system of systems" made of different operative networks. Public, private or hybrid[24]

22 In other words, participation strategies aimed at the creation of new intangible heritage and at the consolidation of the social tissue around it.

23 Here reference is made to the meeting of 13 May 2013 held in Perugia, and cured by Fondazione Perugiassisi 2019, titled *La cultura come sistema di sistemi: cluster, distretti e reti culturali*: http://www.perugia2019.eu/la-cultura-come-sistema-di-sistemi-cluster-distretti-e-reti-culturali/ (*Official website of the event*).

24 Business administration carefully distinguishes between institutes and enterprises. The institute is characterized by its institutional purposes, respect to which the economic dimension is only instrumental. On the contrary, the enterprise is mainly voted at the economic remuneration deriving from private goods. In the specific case of the advanced cultural district, the institute's purpose is connected to the satisfaction of cultural needs of the present and future generations, allowing the formation of new cultural values. The most common juridical forms for the cultural institutes are the public non-economic

institutes and cultural enterprises represent the "nodes" of these networks. They collaborate inside the district in order to (Seddio, 2013):

1. widen their income within the cultural heritage chain: in this regard, it is possible to recognize:
   a. networks with a mono-chain approach – even limited to a single productive phase;
   b. networks with a multi-chain approach which work in several productive sectors acting on the input and output phases of the cultural heritage chain[25].
2. obtain a competitive advantage through integrated activities and functions: in this regard, it is possible to recognize networks based on (Seddio, 2013):
   a. forms of *internal* integration: they base on shared goals, policies, and programs (*political-programmatic integration*) or on a shared management in sites accessibility, public fruition, services to the public and cultural production (*managerial integration*);
   b. forms of *external* integration: they base on an integrated offer of services by the network (*offer integration*) and on a unique image of the network itself *(integration of information, promotion and communication strategies*);
   c. forms of multiple integration: they base on shared strategies for the harmonization of the different levels of the public system of cultural heritage thanks to the collaboration with universities and research centers (*integration of the public chain*) and *with the other close economic chains*.

Furthermore, the cultural systems and the museum districts inside the advanced cultural districts present some particular features respect to the previous networks. They have an internal hierarchy and nodes which are homogeneous for: degree of specialization, content produced, spatial proximity and geographic relevance (regional, national, international) (Alberti and Giusti, 2009).

---

organization and the no-profit institutes. Cultural enterprises, on the other hand, are mainly aimed at the economic remuneration deriving from the production of cultural products and services, even if the production is always accompanied by processes of individual and collective consumption. For this reason, cultural enterprises are defined as "composed holding" (Seddio, 2013).

25 They are important sectors operating behind and before the cultural chain and that in Santagata's theoretical model are excluded (industrial research, cultural tourism, etc.) (Debernardi, 2005; Valentino, 2001).

Considering the characteristics described until now, it is possible to identify the following reticular models of relation within the advanced cultural district (Table 15):

1. the *cultural* network: a regional and national network made by heterogeneous enterprises and cultural institutions which operate within the cultural heritage chain creating horizontal relations and pursuing an internal integration;
2. the (local) cultural system: a national network made by enterprises and cultural institutions belonging to the same territory, but heterogeneous for specialization that operate within the cultural chain, or in its phases, creating hierarchical and horizontal relations, pursuing an internal and external integration;
3. the technological *cluster*: a national and international network constituted by enterprises and cultural institutes homogeneous for specialization but widespread in the territory, which operate in the cultural heritage chain, or in its phases, creating horizontal and hierarchic relation and pursuing internal and external integration;
4. the cultural district: an international network made of enterprises and cultural institutes homogeneous for specialization and belonging to a same territory, that operate in the cultural heritage chain or in its phases creating horizontal and hierarchic relations, pursuing internal, external and multiple integration (only as integration of the public chain);
5. the advanced cultural district[26]: an international network made of enterprises and cultural institutes homogeneous for specialization and belonging to the same territory, that operate in the cultural chain and in the connected production chains, creating vertical, horizontal and transversal relations, pursuing internal, external, and multiple integration.
6. The meta-district: it is an international network made of enterprises and cultural institutes heterogeneous for specialization that do not belong to the same territory and operate in the cultural chain and the connected productive chains, creating horizontal and vertical relations, pursuing internal, external, and multiple integration.

The analysis of these networks confirms the adherence between the relational systems of the advanced cultural district and those of the creative district, with the only exception of the *supra*-local clusters that do not man-

[26] The advanced cultural district is in the list because, being a highly complex "system of systems", it is not possible to exclude that within it operate networks of institutes and cultural enterprises which can be classified as advanced cultural districts.

age to affirm in the cultural field[27]. In addition to this, it evidences the chance for the cultural district to hybridize with other forms of advanced district, as in the tourist district for cultural heritage valorization or in the technological districts for heritage conservation. This confirms the initial hypothesis that the learning process and the knowledge transfer can be explained only if the creative district is considered as a theoretical model (Figure 8).

In synthesis, the advanced cultural district can be defined as a particular form of the advanced industrial district with:

- a dynamic reticular organization;
- systemic interdependencies;
- several forms of integration (vertical, horizontal, and cross-cutting) between the cultural heritage chain and other production chains;
- the ability to generate "hybrid" districts if it works on technological innovation, tourism and food industry.
- economic activities linked to cultural heritage which generate positive relapses even on productions not strictly linked to culture.
- new forms of education, training and learning for the local systems that, nonetheless, only rarely originate *supra*-local clusters.

Figure 8. *The advanced cultural district in Italy: a system of systems*

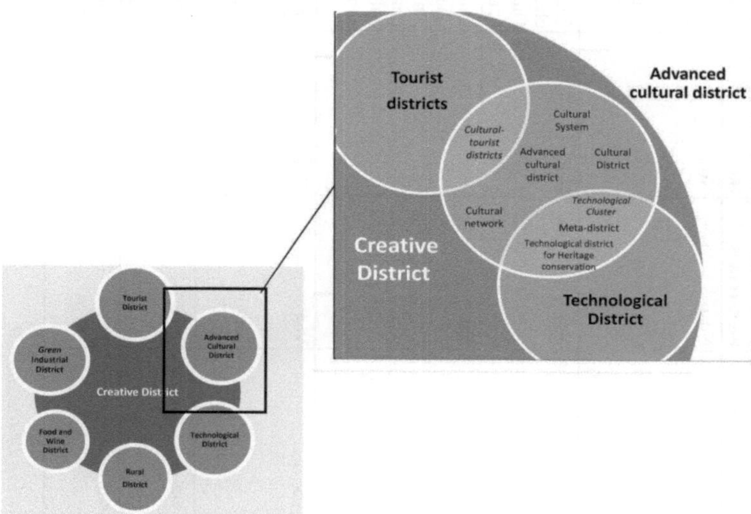

Source: elaboration of the author

27 The cause is a too institutional and institutionalized vision of the cultural district that links the birth of *supra*-local clusters to the adoption of internationalization policies by public administrations.

Table 15. The components of the advanced cultural district: economic-institutional characteristics

Type of network	Spatial proximity of the nodes	Existence of hierarchies	Homogeneity of the nodes	Approach in production	Forms of integration	Geographic relevance	Examples
Cultural network	No Scattered sites within the territory; regional networks, national or international))	No	No	One or more production chains according to the nodes' capacity	Internal	Regional, National	Torino Musei; Musei Piceni, European Network of Coal Mining Museums
(Local) cultural system	Yes	Yes	No	One production chain. Only the cultural heritage chain	Internal External	National	Sistema Culturale Rhodense (Yes.Cur.Rho) Milan (IT)
Technological Cluster	No	Yes	Yes	One production chain. Only some phases of the cultural heritage chain	Internal External	National	DiT-BECS, Florence (IT)
Cultural District	Yes	Yes	Yes	One production chain. Only the cultural heritage chain	Internal External Multiple (harmonization of the public sector)	International	Cultural Districts by Fondazione Cariplo, Lumbardy (IT)
Distretto culturale evoluto	Yes	Yes	Yes	Plurality of production chains	Multiple	International	Advanced Cultural Distric fo Marche Region (IT)
Meta-distretto	No	Yes	No	Plurality of production chains	Internal External Multiple	International	Metadistretto Veneto dei Beni Culturali e Ambientali, Veneto Region (IT).

Source: elaboration of the author

The Advanced Cultural District in Regional Planning

Instruments and Methodologies

The eco-systemic approach of the studies on industrial organisation has allowed a better understanding of the creative spatial policies. The interest in urban regeneration and local development has encouraged their spread in Europe, especially thanks to the EU programming where the creative spatial policies are a starting point for the construction of advanced cultural districts. In order to identify the social, economic and spatial dimensions of these policies new analytic tools have been developed with the support of several disciplines (Table 16).

The following paragraphs provide an overview on these instruments and on their applications in the construction of advanced cultural districts pointing out the local development factors considered in this process.

Table 16. The three dimensions of creative spatial policies according to the eco-systemic approach

	Topic	Actors	Resources
Social dimension	Individuals and creative systems	– creative talents – production organisations (networks, systems, clusters, districts, meta-districts, etc.)	Human resources: relational and communication skills
Economic dimension	Cultural and creative economy intended as industrial sector	– Cultural and creative companies	Economic resources: cultural and creative products and services
Spatial dimension	Strategies for the protection of the built environment, linked to urban regeneration and local development	– public bodies – cultural and creative companies (including designers and planners) – non-profit organisations – public-private partnerships (PPPs)	Local resources: – cultural and natural sites – centres of cultural production and consumption

Source: elaboration by the author

5.1 The existent instruments and their effectiveness

5.1.1 The economic survey: statistical indicators, analysis of the relationships between companies and production chains

The first step towards the construction of a district is a statistical analysis on the local cultural and creative production, on the competitiveness of the district in the global, European, national and regional markets (city ranking and benchmarking).

The relating literature provides various panels of indicators[1]. The choice depends on the importance given to cultural heritage according to the approach adopted (econo-centric, culture-centric or integrated) (Table 17). In Italy, for example, statistic reports generally focus on creative industries and are characterized by a culture-based or integrated approach. The latter, in particular, is the distinguishing feature of the most recent reports based on the analytical model of "cultural platforms" (Table 18).

Table 17. International statistic frameworks

Cultural consumption (Florida's 3Ts)	
	– UN Creativity Index (Florida *et al.*, 2011) – EU Creativity Index (KEA, 2009; Florida and Tinagli, 2004)
Cultural production (cultural and creative industries)	
Econo-centric approach	– International Intellectual Property Alliance, USA (2006) – EUROSTAT (2007) – KEA European Affairs (2006, 2012, 2013) – DCMS,UK (2007) – DEPS, FR (2006)
Culture-centric approach	– American for the Arts, USA (2007) – United Nations (2008, 2010, 2013) – UNESCO framework for cultural statistics (2009)
Integrated approach	– Municipal Cultural Resources Framework, Canada – (MPCI, 2011) – Creative City Index, Charles Landry, UK (www.charleslandry.com) – INTELI (2011)

Source: elaboration of the author

[1] The statistical models for the cultural and creative sector are detailed in chapters 2, 3 and 4.

Table 18. Italian statistic frameworks

Cultural consumption (Florida's 3Ts)	
	– Italian Creativity Index (Florida and Tinagli, 2005) – Report on the economy of culture in Italy 1990-2000
Cultural production (cultural and creative industries)	
Econo-centric approach	– Report "La città dei creativi" (ANCE, 2005) – Statistical classification of the economic activities: Ateco 2007 built on NACE Rev. 2 (ISTAT, 2009)
Culture-centric approach	– Cultural Profitability Index (Santagata, 2007) – Report "Le attività economiche collegate alla valorizzazione del patrimonio cultural" (MiBAC e Istituto Tagliacarne, 2007) – Report "Il sistema economico integrato dei beni culturali" (Unioncamere and Istituto Tagliacarne, 2009) – White Paper on Creativity (MiBAC, 2009) – GAI, Italian Creativity Index (Cicerchia, 2010, 2013)
Integrated approach	– 4T's Model (Carta, 2009) – Report "La valutazione sulla creatività nelle regioni italiane" (Pini and Rinaldi, 2010) – Report "L'Italia che verrà. Industria culturale, made in Italy e territori" (Unioncamere and Symbola, 2011, 2012, 2013, 2014) – IX Report of Federculture "Una strategia per la cultura una strategia per il paese (Grossi, 2013) – Book "La nuova occasione. Città e valorizzazione delle risorse locali" (Casavola and Trigilia, 2013) – BES Report (ISTAT, 2013, 2014)

Source: elaboration of the author

Once the statistic survey has identified the best productions, the sectors in crisis, the best economic performances and the size of companies of the future district, it is necessary to investigate the relationships existing between the creative talents and the creative systems (networks or clusters). The techniques used in this case are: the analysis of the inter-organisational relationships (*social network analysis*), the "subject-type of relationship" matrix, the cluster analysis and the matrix of the inter-dependence among sectors.

The analysis of inter-organisational relationships, or *social network analysis* (SNA), helps to identify the key actors of the future district, whether they are individuals, companies or public bodies (Edizel, 2013). It consists of the following steps (Lazzeretti, 2012):

1. Sampling of the projects and of the actors involved in the local creative spatial policies in a fixed period of time;
2. Creation of a database of the projects: choice of the descriptive parameters for each project (title, funding body, year, coordinator,

number of partners, key elements, outcomes), filling of the database with the data of local projects;

3. Creation of a project-actor database: choice of the descriptive parameters for each actor[2] (name, location, type, area of competence, financial contribution received, etc.) and filling of the database with the data of the local actors;
4. Analysis of the statistical distribution of the project actors as well as of the financial resources for each selected parameter (distribution of the actors by location, type, etc.; distribution of the financial resources by the actors' location and type, etc.);
5. Creation of the affiliation matrix: a squared binary matrix where the rows identify the actors, the columns identifies the projects (events) and the cells indicate the involvement of each actor in a certain project (1) or, conversely, their absence (0);
6. Creation of co-membership matrices: square binary matrices where the rows and columns indicate the actors and the cells show the involvement of each couple of actors in a certain project (1) or, conversely, its absence (0);
7. Creation of *graphs* by means of a SNA software which operates on the co-membership matrices according to the following indicators:
   - network density: the ratio between actual and potential relational ties (the latter referring to the ties that might occur if each actor were connected to all the other actors);
   - centrality degree: the number of direct ties of an actor (it depends on the degree of co-memberships in the projects);
   - betweenness centrality degree: when an actor connects two non-adjacent nodes on the network acting as a bridge;
   - centralisation degree: number of direct links of a cluster of actors (it depends on the degree of co-memberships in the projects);
   - betweenness centralisation degree: when a cluster of actors acts as a bridge with respect to other non-adjacent nodes on the network (it depends on their willingness to establish reiterated co-memberships with different actors).

The resulting graph provides a two-dimensional representation of the ties existing among the actors involved in local creative spatial policies. The graph can convey important information about the position of each person, company or public body within the network, as well as about the morphology

[2] These parameters stress the simultaneous involvement of the same actor in one or more projects, thus facilitating the following steps of the SNA and the identification of the key actors for the activation of local resources in view of the construction of the future district.

and the overall functioning of the latter. For example, in Tuscany, the analysis of the inter-organizational relationships within the industry of technologies for heritage conservation highlighted the presence of a cluster and promoted the establishment a thematic district, named DiT-BeCS (Figure 9). Moreover, the graphs identify the core of the cluster – responsible for the local buzz – and the national, European and international actors engaged – responsible for the global pipelines (Bathelt *et al.*, 2004) (Figure 9).

Figure 9. *The cluster of laser technologies used for the conservation of the cultural heritage in Florence.*

Compartecipazione ad almeno due progetti (c>2):

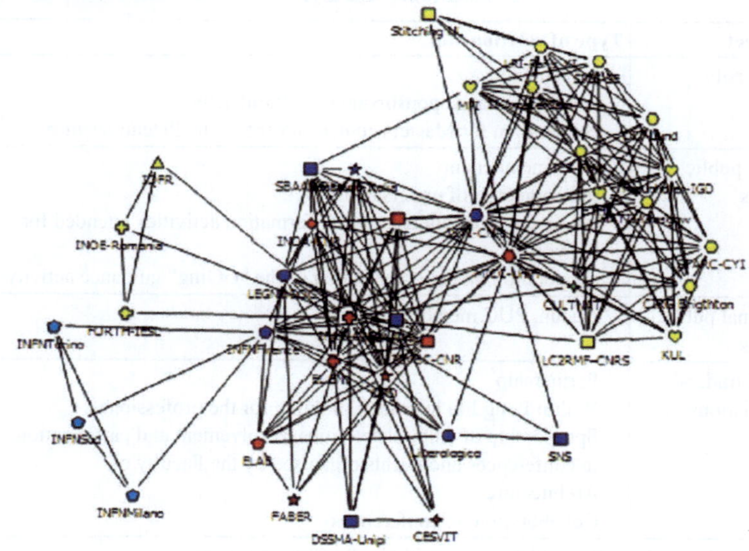

Compartecipazione ad almeno quattro progetti (c>4):

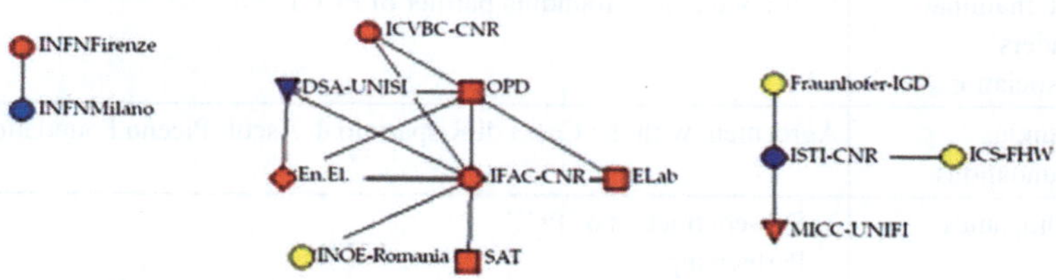

Co-membership in at least two projects (c>2)
Co-membership in at least four projects (c>4)
Note: the colour shows the location of the nodes: red for Florence, violet for Tuscany, blue for the remaining part of Italy and yellow for Europe.

Source: Lazzeretti (2012)

After the identification of the most significant local clusters, it is possible to determine the formal and contractual nature of the relationships which tie the actors of the future district. The tool used is the "subject-type of relationship" matrix, i.e. a grid which describes the legal, administrative and contractual instruments used by each actor (*subject*) to collaborate with other public or private actors in a project. Emblematic is the matrix created for the "Feasibility study for the design of an advanced cultural district in the Province of Ascoli Piceno (IT)" (see Figure 10).

Figure 10. *The matrix applied to the relational system of the Piceno University Consortium – PUC (IT)*

Subject	**Type of relationship**
University	– agreements – cooperation and sponsorship of cultural activities – cooperation for Masters courses hosted in the Piceno territory
Local public bodies	– PUC membership – memorandum of understanding – cooperation on guidance and information activities intended for the youth – cooperation on the organisation of the "GOing" guidance activity
National public bodies	– Ancun, PUC member
Local traders' associations	– Partnership – Within Long life learning – training for the professionals – Sponsorship of cultural activities, involvement and participation in conferences and events organised by the Faculty of Architecture – Collaboration on conferences
National traders' associations	ADI: Collaboration with the Institute of Design
International traders' associations	Erill Foundation (founding partner of PUC)
Banking foundations	Agreement with the Cassa di Risparmio di Ascoli Piceno Foundation
Companies	– Sub-contractors of PUC – Partnerships – Subscription to the Campus Card, a charter for services established by the Piceno University Consortium which allows students to buy products and services at discounted prices – Leonardo Project.

Source: Sacco (2007)

Another analytic technique used to qualify further the economic actors of the cultural and creative sector in the future district is the cluster analysis. Compared to the SNA, the cluster analysis focuses on companies by investigating the connections with the other economic sectors and with the other production chains from a statistical point of view. The purpose is the recognition of economies of scale which could allow the district to compete globally. Various tools can be applied, however, all of them include the following steps (Maggino, 2005, pp. 1-22):

1. Selection of companies which are the object of the cluster analysis (*elements*) to be investigated according to their relationships:
   a. in case of relationships between the companies indicated by the SNA, only the respective statistical areas (NACE classification) shall be considered;
   b. in case of relationships between companies in the cultural and creative sector, the statistical model applied to the city ranking and benchmarking[3] shall be adopted;
   c. in case of relationships between all the companies of the future district, the official reference classifications (ISTAT, Confindustria, Unioncamere, Chambers of Commerce, etc.) shall be used.
2. Identification of the statistical variables underlying the clustering[4];
3. Creation of the data matrix, whose rows indicate the elements, the columns identify the variables and the cells show the performance of each element with respect to a certain variable (numerical value);
4. Standardization of the variables, when it comes to cluster the elements (rows of the data matrix) and the variables are different from one another with respect to the values used (measurement scales), or standardization of the profiles, when the target is to cluster variables (columns of the data matrix) and the elements used are different from one another in the profile values;

3 In the Italian case, the references for the identification of the elements are the ATECO 2007 classification and the statistical framework provided by the White Paper on Creativity (MiBAC, 2009).

4 For example, the method used by the ISTAT for the mapping of the industrial districts, starting from the local employment areas (SLLs), takes into account the following variables: percentage of total manufacturing workers with respect to the total number of workers; percentage of the manufacturing workers of SMEs compared to the total number of manufacturing workers (SMEs <250 workers); percentage of workers by sector with respect to the total number of manufacturing workers; check for the actual presence of 50 % of specialized workers at SMEs (Buscema *et al.*, 2012).

5. Setting of the similarity or distance index: is assesses the element-element, element-cluster and cluster-cluster ties in terms of similarity or distance;
6. Definition of the clustering strategy, including the number of final clusters and the method adopted for their creation;
7. Choice of the classification algorithm (clustering technique): based on the clustering strategy (step 5), it is possible to choose the algorithm which will act on the data matrix (or on the dissimilarity matrix) as to obtain a dissimilarity matrix[5];
8. The clustering through a specific software: starting from the data matrix, the proximity index adopted and the algorithm chosen, the software creates *n* dissimilarity matrices until the programmed number of clusters is achieved;
9. Interpretation and validation of the results, also though diagrams.

The cluster analysis identifies the most important companies from both a production and a local point of view, since the definition of the variables and the distance index may include indicators relating to geographical position, infrastructure and availability of public services.

The "Feasibility study for the design of an advanced cultural district in the Province of Ascoli Piceno (IT)" contains an interesting cluster analysis based on the concept of "multi-location company network", which investigates the relationships among all the businesses of the future district according to five variables of reference[6] (Sacco, 2007). This study highlights the spatial distribution of industrial centres and hubs in Marche region through the identification of eleven homogeneous clusters. They are very similar in the manufacturing activity, technological endowment, social structure of the labour supply system and transport infrastructures (Figure 11). Conversely, the "Reggio Emilia Città Creativa" research project provides an example of cluster analysis applied to the creative and cultural sector according to the statistical models used in city ranking, with the purpose of identifying a local cluster on new media (Figure 12).

---

5 Many clustering strategies are iterative, i.e. they repeat the same step several times until the desired number of clusters is achieved. In the first step, the algorithm acts on the data matrix by assessing the tie between two elements, thus creating a distance matrix. Conversely, from the second step on, the algorithm acts on the distance matrix considering the element-cluster or cluster-cluster tie, which is called dissimilarity, and creates a "dissimilarity matrix."

6 The analysis has been performed on the data provided by Confindustria Marche.

Figure 11. *Cluster analysis: the feasibility study for an advanced cultural district in the Province of Ascoli Piceno (IT)*

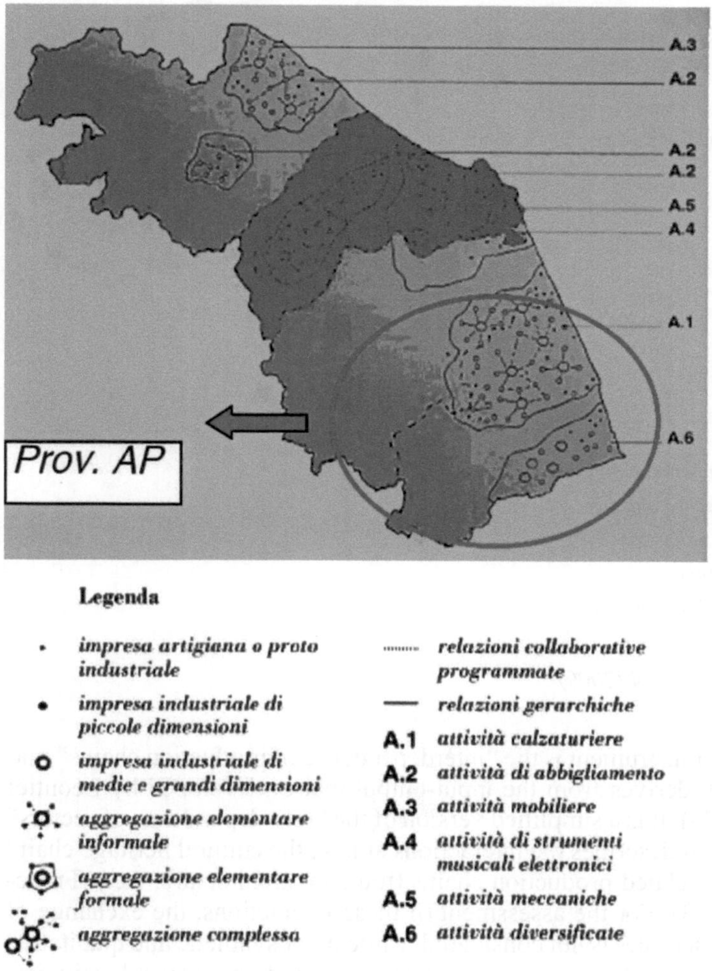

Craft or proto-industrial firm
Small industrial company
Medium and large industrial company
Informal basic cluster
Formal basic cluster
Complex cluster
Planned collaborative relationships
Hierarchical relationships
Footwear
Clothing
Furniture production
Electronic music instruments
Mechanical engineering
Multi-sector companies

Source: Sacco (2007)

Figure 12. *Cluster analysis: "Reggio Emilia Città Creativa" project*

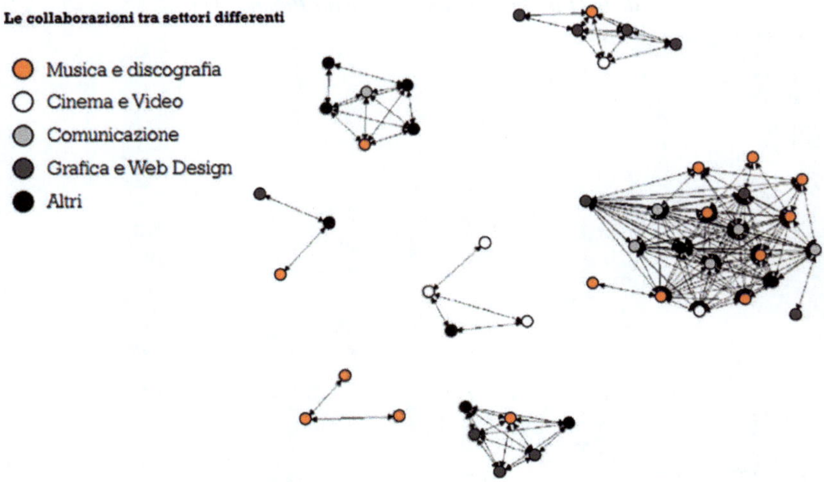

Cooperation between different sectors
Music and record industry
Cinema and video
Communication
Graphics and web design
Other

Source: Montanari (2009)

Another instrument is the "interdependence of production chains" matrix, which derives from the input-output matrix developed by Leontief (1951, 1986). It is a simplified version of the "interdependence of sectors" matrix and it describes the interactions among the cultural heritage chain[7] and the co-related production chains (tourism, architecture, etc.) (Francesconi, 2013). For the assessment of these interactions, the exchange of goods and services is not considered, while other synthetic and qualitative parameters are taken into account such as: shared projects and activities, exchange of information and knowledge, sharing of managerial and organizational practices as well as of resources and competences (including the technological ones).

[7] Cultural heritage chain is intended herein as a series of activities which characterize the "life cycle" of a cultural product: protection, preservation, enhancement, management, promotion and fruition. This definition was introduced by Unioncamere and the Tagliacarne Institute (2009) and is based on the chain model suggested by Walter Santagata (2007), which consists of following steps: selection of creative artists and actors, development of ideas, production, distribution and fruition.

The application of the matrix must be preceded by a study of the production chains co-related to the cultural heritage. They can be identified and described through the clustering of the production companies by means of:

- a social network analysis
- a statistic analysis of local programming on culture according to the institutional classification systems (for Italy, ISTAT, Confindustria, Unioncamere)
- a panel of indicators, such as those used in city ranking[8] (Table 19).

The techniques explained so far provide important information on the economic relationships arisen from creative spatial policies: the turnover of creative and cultural sector, the key actors (local buzz and global pipelines), the ties among sectors, the link with the other production chains (clustering and interdependence), the nature of these relationships (legal and contractual form). However, the connection between the creative talents and the creative systems is expressed only in economic terms disregarding the geographical data which are considered only indirectly, as in the case of Florence or Marche region. Therefore, the economic analyses must be integrated by a spatial survey.

Table 19. Interdependence matrix

Cultural heritage		**Production chains**						
		Fashion	**Industrial design and handicraft**	**Food and wine industry**	**Software**	**TV and radio**	**Advertising**	**Cinema**
Cultural chain	**Protection**							
	Preservation							
	Enhancement							
	Management							
	Promotion							
	Fruition							

Source: elaboration by the author on Francesconi (2013)

8 In Italy, the reference is the panel of indicators outlined in the White Paper on Creativity by MiBAC (2009), curated by Walter Santagata.

5.1.2 The spatial survey: distribution (mapping) and context analysis

The geographical proximity intended as the spatial concentration of companies in a defined area, has been regarded as a favourable factor for districts since the time of Alfred Marshall. To this end, the techniques of urban planning are often used in the feasibility studies of district. They are used to investigate the spatial distribution of firms and the influence of the context factors on the production process. For instance, the distribution analysis reproduces the information on cultural and creative industries (CCIs) on a cartographic support and provides information on the geographical concentration of firms.

The mapping refers to the classification system adopted by the national institutes of statistic for the cultural and creative sector and to the Company Register at the local Chamber of Commerce. Each company must be classified through: the NACE code; the creative/cultural sector of belonging; the legal address; the individual or collective nature of the company. The location is reported on a geo-referenced digital map by CAD and GIS softwares. The assessment of the distribution and concentration of companies relies on the selected sampling area (census area, administrative units, etc.) along with the following parameters:

a) number of creative and cultural companies with the same address: the parameter is put in relationship with the population density, the price per square meter of the offices, the year of construction and of renovation of the building (Figure 13);
b) addresses which host a number of companies above the average of the sample area: these addresses are called hotspots (Figure 14);
c) number of creative and cultural companies per thousand inhabitants;
d) number of creative and cultural companies with respect to the total number of companies.

The analysis is conducted for each parameter on the entire sample of companies (aggregate data) as well as on the companies operating in each creative field (disaggregated data). In both cases, the data are further processed by considering collective and then individual companies.

Once the mapping of companies is completed, the analysis of the context factors, i.e. the study of the context-related issues able to influence the development of creative companies in a territory, is carried out. It relies on an in-depth investigation and in a detailed study. The in-depth investigation identifies the environmental, managerial and design factors within the creative and cultural sector which exercise a stronger impact on the cultural heritage chain as well as on the co-related ones[9].

Figure 13. *Spatial distribution of creative companies with respect to the population density in Ljubljana (Slovenia)*

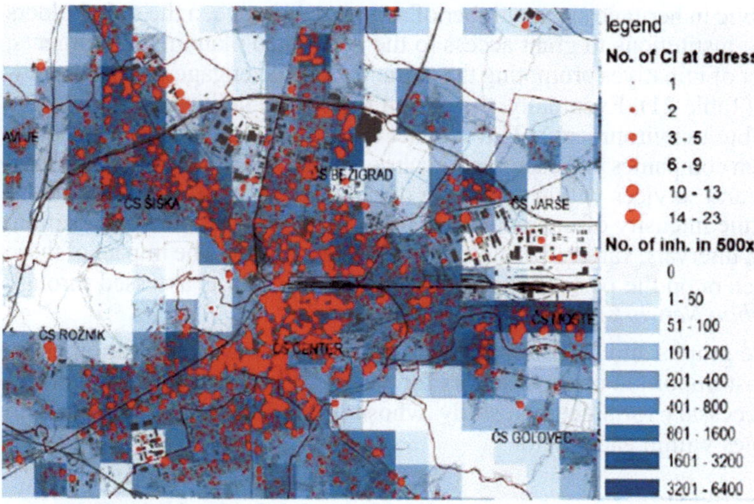

Source: RRA LUR (2011)

Figure 14. *Spatial distribution of creative companies in relationship to the year of construction of the building, with creative hotspots (black circles)*

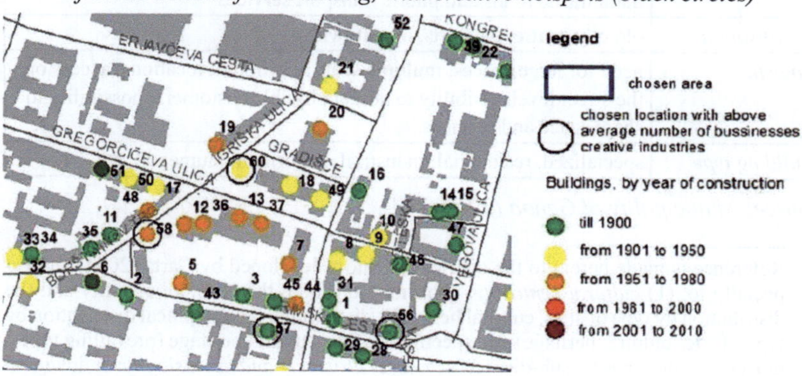

Source: RRA LUR (2011)

9 The MiBAC model includes: fashion, design and handicraft, food and wine industry, software, advertising, cinema, TV and radio.

The context factors are defined as peculiar aspects which influence the choice of the production site and the establishment of the company[10] (Table 20). Conversely, the management factors are intended as socio-economic aspects of the context able to affect the establishment and management of the company. They include: extent of heritage preservation, enhancement and promotion; training and scientific research; availability of services for the public in heritage sites; number of agreements between the stakeholders and the institutions to grant access to the negotiated planning instruments; number of initiatives promoting the participation and engagement of the citizens[11] (Table 21). Eventually, the design factors are defined as the features of the built environment able to impact on the opportunities of interaction between companies and customers such as, for example, the number of companies and services within a walking distance; the spatial qualities of these paths, the intensity of interactions in relation to building's shape, position, shelter, intervals, variety, permeability and accessibility. The impact of these parameters on the performance of creative companies is assessed through the *Urban Network Analysis* (UNA) (Figure 15).

The graphs and maps resulting from the in-depth investigation show the most important context factors for each sector, providing the information necessary for a detailed study, whose purpose is to identify the main variables within each factor.

Table 20. Context factors: classification and definition in the *Genoa Creative Cities* project

Factors	Definition
Accessibility	direct access through large means of transport or proximity to the main traffic roads and public transport services
Localisation	old city centre, outskirts, downtown
Logistics	need for large spaces, multiple seats in different locations, access on the street level, visibility to attract possible customers; possibility to share space and services.
Building type	specialized, residential, industrial, historic; monumental, etc.

Source: Municipality of Genoa (2013)

10 Reference is made herein to the district indicators developed by Carta (2004) and especially to: (1) *cultural centrality*: concentration (both the absolute quantity and the distribution by type of sites, cultural heritage), distribution (geographical distribution by type of sites cultural heritage) and specificity of the cultural heritage (prevailing types, most common types by sub-district areas); (2) *proximity and accessibility*: endowment and efficiency of the necessary infrastructure (opening times and days for the sites; infrastructure and public transport; access signs; price; information technologies for the access to multimedia contents) (Francesconi, 2013).

11 Reference is made herein to the indicators developed by Carta (2004), Sacco and Ferilli (2006) and by Francesconi (2013).

Figure 15. *Design factors: the* Urban Network Analysis

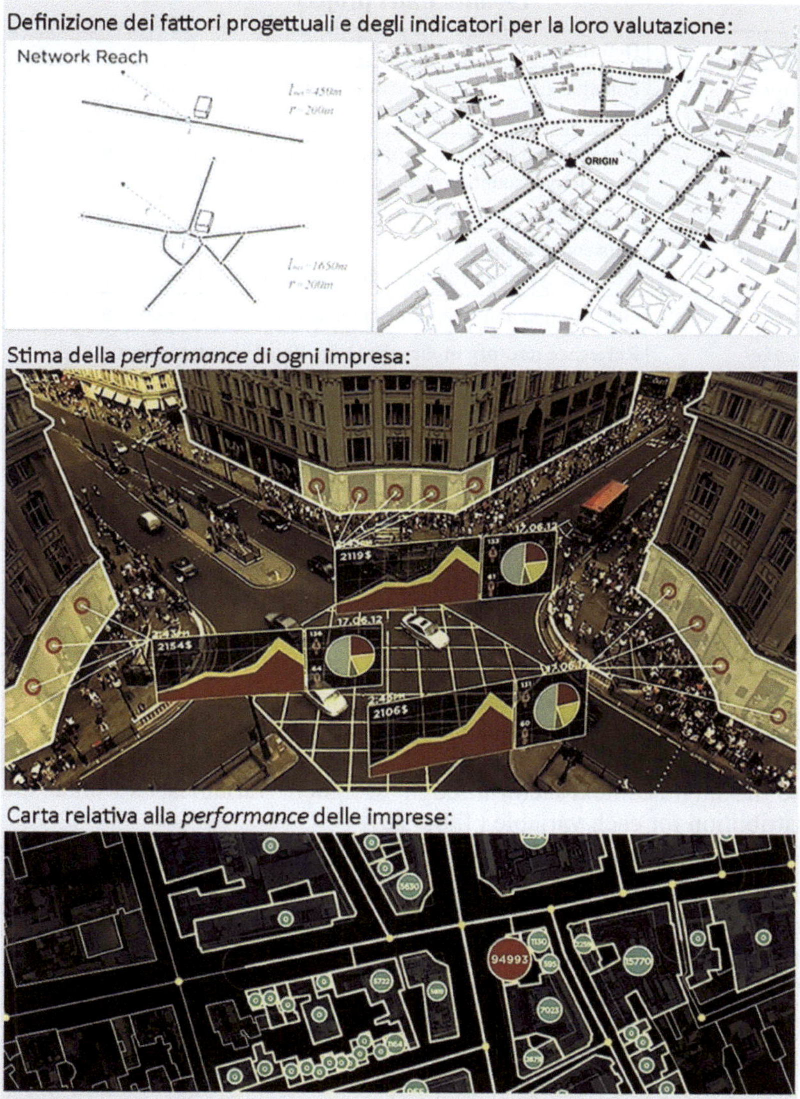

Definition of the design factors as well as of the indicators for their assessment
Performance assessment for each company
Map relating to the performance of the companies

Source: City Form Lab (2014)

Table 21. Management factors: classification and definition in the *Genoa Creative Cities* project

Factors	**Definitions**
Target	Potential customers (public bodies, individual business and companies).
Satellite activities	Presence and localization of satellite activities. The variables taken into account show the reference position of the satellite activities (local or extraterritorial).
Network	Existence of a network; Network typologies: weak, strong or industry-related.
Structure	Organization and size of the company (individual or collective).
Market	Reference territory of the customers (local or extraterritorial).
Regional policies	Presence of support measures already in force in the area (incentives, services, territorial brands).

Source: Municipality of Genoa (2013)

For example, within the *Creative Cities* project of Genoa, the in-depth investigation consider accessibility as the most important context factor and, therefore, the detailed study has focused on variables indicative of the topography and polycentricism of the settlements. The percentage of the internal distribution of the variables for each factor results from the use of a range of survey instruments: from a matrix of "internal consistency" to interviews with key-stakeholders. These assessments provided a detailed maps showing the links between the distribution of creative companies and the single context factor through "weighed" variables, as well as the distribution for each variable (Table 22).

Again with reference to Genoa, the interviews with the key stakeholders of the creative sector highlighted that, as for the accessibility factor, the leading variable was the proximity to main roads as well as to public transport. This led to the development of new "accessibility" maps in addition to the detailed ones referring to the two specific variables.

Table 22. Matrix of "internal consistency" for the *Creative Cities* project in Ljubljana (Slovenia)

Sectors of Creative activities	Population density (critical mass)	Ambiance component (environmental, aesthetic attraction of location)	Price of real estate/ rent (age of building, renewal etc.)	Importance of infrastructure (size of space, mobility, closeness to a transport network)	Accessibility of customers (markets, micro locations)	a) Gradual accumulation of creative production ("effect of a snow ball")	b) Exponential, rapid accumulation of creative production (efect of coincidence, "open door", suddenly create favourable conditions for CI)
Advertising	++	++	++	+++	+	++	+
Architecture	+++	+++	++	++	+++	++	+++
Creative production	+++	++	+++	++	++	+	+++
Retail sale of arts and antiques	+++	+++	++	o	+++	++	o
Design	++	++	+++	+	+	+	++
Institutions of culture and creativeness	++	+++	+	+	+++	+	+
Music and visual production	+++	++	++	+	++	++	++
Publishing	++	+	+	o	+	+	+
Radio and TV production	+	o	o	+	o	o	++
Computer programming	++	+	++	++	++	++	+++
Video, film and photo production	++	+	++	++	+	+	+++

Note. For each variable, the best performing sectors are indicated with "+++", while the poor-quality ones bear a "++" mark, and the worst ones correspond to a "+". The indifference or lack of influence of the sector with respect to the variable considered is expressed with "o".

Source: elaboration by the author on RRA LUR (2011)

5.1.3 The sociological survey: the appraisal of competencies and resources

The advanced cultural district is different from the Anglo-Saxon cultural district due to the importance attributed to the intangible heritage of knowledge, beliefs and values which characterises the local production systems as relational networks. Therefore, the analysis of creative spatial policies is often supported by an appraisal of human, cultural and social capitals (the "intangibles" resources of the district).

In training and re-motivational activities the *competence appraisal* requires a systematic analysis of the individual abilities and is carried out by means of tests and self-assessment grids (Evangelista, 2007, 2008). Conversely, during the construction of a district, the skill assessment requires a collective in-depth reflection and an open discussion among the actors of the creative spatial policies. This happens through structured questionnaires, like the one introduced by Francesconi (2013). All the actors are asked to answer the questions and, later, to confront each other by focusing on the resources and skills actually available, as well as on the strengths and weaknesses of the territory of belonging (Table 23).

Table 23. Example of an appraisal of competencies and resources for a cultural district

Factors	Importance	Relating power	Description
	On a 1-10 scale	*On a 1-10 scale*	
Skills/competences			
Management of interdependences and creation of synergies among production chains			
Management and promotion of relationships among the actors of a network			
Knowledge management			
Management of (inter)organisational learning processes			
Management of inter-dependence and creation of synergies among chains			
System thinking			
Combination of resources and competences			
Inter-institutional project management			
Brokering, mediation and arbitrage services			
Activation and management of cultural networks and systems			
Development and management of information systems to support the network community			
Social and economic accountability			
Strategic monitoring and control			
Benchmarking and benchlearning			
Social innovation and entrepreneurship			
Triggering and support, institutional change, dynamic capability			
Re-combination of resources and competences			

Factors	**Importance**	**Relating power**	**Description**
	On a 1-10 scale	*On a 1-10 scale*	
Resources (*continues*)			
Tangible cultural heritage			
Intangible cultural heritage			
Landscape heritage			
Human capital			
Social capital and network			
Intangible territorial heritage			
Infrastructure territorial heritage (territorial structure)			
Administrative integration/interdependence resources			

Source: Francesconi (2013)

The actors involved in the skill assessment are selected through *bench-marking* statistical analysis, *social network analysis* or – in the absence of any preliminary analysis – through the *snowball sampling*[12] system.

The appraisal of competencies and resources is a bottom-up procedure which rounds off the district building process by providing feedback about the "top-down" economic and spatial analysis of the territory carried out by the commitments, or, alternatively, by a working group appointed by them for the draft of the feasibility study.

5.2 Building an advanced cultural district: methodological and instrumental aspects

In Italy, the construction of an advanced cultural district is a long and complex process with an unpredictable outcome. It requires a great effort by the sponsoring body – generally made up of regional administrations or banking foundations.

In the last fifteen years, the regional tenders for the construction of advanced cultural districts supported numerous feasibility studies but they

[12] See Goodman, L.A. (1961). Snowball sampling, in *Annales of Mathematical Statistics*, 32(1), pp. 148-170

culminated with the recognition of just two districts per region[13], expect for the Lombardy region where six cultural districts have been created by Fondazione Cariplo.

The Lombard districts of Fondazione Cariplo have benefited from the technical support of Fondazione Cariplo, which has organized a training programme and provided the potential districts with a team of experts and with guidelines for draft of the feasibility studies. The results, the procedural innovations and the affinity with the regional policies on districts turn the Lombard experience into a reference model in the construction of cultural districts. On this basis, we make reference to the methodology of Fondazione Cariplo to identify the key steps in the construction of an advanced cultural district.

5.2.1 The procedure

The first step consists of an in-depth survey which aims at identifying the potential districts in the territory of reference. This research is supported by the sponsoring body, which may either act directly or appoint external advisors. As already explained, the sponsoring body can be public, private or mixed (regional administrations, banking foundations, a consortium of local bodies and companies, etc.).

The identification of the potential districts requires the use of statistic indicators on: cultural and environmental heritage; population; training and research activities, synergies with the economic system; ability to develop projects and alliances; administrative systems; management skills (Della Torre, 2006).

The assessment of each factor relies on the data provided by official sources (ISTAT, Confindustria, local bodies, chambers of commerce, etc.) as well as on the policy documents on the creative spatial policies[14] produced within a period of time (maximum ten years). In the socio-economic survey, the identification of a statistical model for the analysis of the cultural and creative sector and the recognition of local chains play a crucial role. Based on the outcome of the in-depth analysis, the sponsoring body issues a call for tenders which defines the criteria for the assessment of the project proposals.

The project proposal consist of a pre-feasibility study for the construction of a cultural district at a sub-regional scale. The proposal content is described in the tender and it must include (Fondazione Cariplo, 2007):

13 In view of a great number of "cultural districts only on papers", as stressed by Hinna and Seddio (2013). In fact, public administrations were expected to keep about one hundred pre-feasibility or feasibility studies which unfortunately were not followed by any testing, monitoring and assessment.

14 Here reference is made to the classification performed by Sager (2011).

1. Description of the district: it provides information on the territorial, socio-economic, institutional and legal context; the cultural and tourist supply and demand; it includes an in-depth analysis of the development needs, final targets, strengths and weaknesses of the district area;
2. Document of Technical feasibility: it provides information on all the interventions, the implementing bodies and the technical aspects relating to the implementation (project progress, timing, description, location, importance and urgency of the interventions on valuable real estate properties, etc.);
3. Document of Financial feasibility: it provides information on the economic investments necessary for the development of the district; it includes a financial security plan attesting to the reliability of the co-financing commitment and to the district management budget; the economic and social feasibility analysis (cost-benefit analyses over a period of 5-10 years);
4. Document of Administrative-legal feasibility: it provides information on the partnerships to be activated (leading body, partners, involvement level, role, operational and financial commitments of each party), agreements signed by the various bodies, consistency with the territory, the targets and the budget of the district.

After the selection, the sponsoring body approves the winning pre-feasibility studies and asks the competitors to formalize the working group and validate the "feasibility study monitoring plan" required by the sponsoring body (Chiavarino *et al.*, 2013). This plan consists of a timeline which defines the deadlines for the submission of the periodic reports on the progress of the feasibility study, the dates of the regular meetings with the advisory board as well as of the plenary meetings with the competitors.

The Advisory Board is made up of external experts chosen by the sponsoring body and belonging to the technical-scientific committee which has selected the pre-feasibility studies. The Advisory Board has to examine the progress report presented by the working groups, highlight any critical aspects and delays, ask for further information about specific choices and procedures and suggest the proper communication of the information so as to grant the transparency of the whole process. The Advisory Board performs its coordination and control tasks through regular meetings with each working group as well as through plenary meetings with all competitors. In the first case, an update on the current situation is provided, any doubt, negotiation issue and omission is clarified, and the internal relational dynamics are developed in a positive way (avoiding condescension or disputes between the committent and the advisors of each working group).

The representatives of the sponsoring body who take part in these meetings produce a written evaluation and send it to the Technical Committee and to each working group. This document contains an assessment of the periodical reports, neglecting the additional information verbally delivered by the working group during the meetings. This allows any useless and residual communication and any information in conflict with the orientation of the selection process to be overlooked.

During the plenary meetings, which are technical, the representatives of the sponsoring body and the technical committee share opinions and discuss with the working groups about the draft and the selection of the feasibility studies: general targets, contents of the guidelines, financing mechanisms, criteria for the assessment of the proposals (Chiavarino *et al.*, 2013).

An additional support to the working groups are the guidelines provided by the sponsoring body. They are drawn up by the technical committee in the form of operating manual. They contain instructions about the minimum contents of the feasibility study, together with the goals and size of each analysis; however, they avoid any provision concerning the methodologies to be applied to collect and process the data (Cammelli *et al.*, 2007).

At the end of the period defined by the sponsoring body for the draft of the feasibility studies, the assessment of the proposals submitted by the working groups can begin. It relies on:

1. the technical evolution form by the Advisory Board which states the quality of each feasibility study (Table 24);
2. the written opinion of the representatives of the sponsoring body: it is based on the feasibility study as well as on the reports provided by the technical committee and it includes a proposal of budget.

Once these documents are analysed, the sponsoring body selects the feasibility studies to be funded and approves the budget allocations for the winning proposals.

After the approval of the feasibility studies, the sponsoring body finances the start-up of districts. It supervises the implementation of the projects included in the management plan by means of four instruments: agreements, constraints, a monitoring system for the work in progress and an *ex-post* assessment (Canino *et al.*, 2013).

Table 24. Assessment criteria for the reports provided by the technical committee

Assessment criterion	**Parameters considered**
Effectiveness of the feasibility study	– probability to built the district; – process adopted for the definition of the district proposal; – problems which may undermine the creation of the district (risk factors); – effectiveness of assessment indicators: complete contents, clear procedures, consistent methodology.
Quality of the cultural district	– possibility of integration into the local cultural system and connection to the local production context; – suggested governance approach (especially in relation to the stability and institutional legitimacy of the district initiative); – deepness and completeness of the financial, economic and management plan; – resolution documents and financial commitment (actual allocation of funds) by the project partners.

Source: elaboration by the author on Sotti (2013) and Chiavarino et al. (2013)

The sponsoring body sings an agreement with the committent (or leading partner) of each working group awarded. Upon the signature the working group and the other stakeholders involved recognise the sponsoring body as the project leader and accept the rules and the intervention strategies set out in the same agreement (interventions, implementing bodies, financial-economic plans)[15]. Within the agreement, the sponsoring body may divide the interventions into several categories based on their relevance or on their nature. It may also introduce a monitoring and reporting system with operating and accounting constraints more or less stringent depending on the risks[16].

15 The agreement is signed by the sponsoring body and the leading partner engaged in the promotion of the cultural district.

16 The relevance is assessed according to the effects resulting from the failure to complete the works in the district: (1) crucial actions: they are the strategic core of the district, their absence affects the construction of the district, the risk to the sponsoring body is high; (2) highly-relevant actions: they are important but include some uncertainty due to uncontrollable external factors; they can be replaced with actions with the same relevance or economic value and the validity of the district project is not impaired, the risk to the sponsoring body is limited; (3) Relevant actions: they are important and feature no special problem of implementation, in case of their absence, a single allocation can be revoked without changing the whole district project, the risk to the sponsoring body is low; (4) important actions: they are similar to the previous ones but, in this case, the shift of the financing contribution from one action to another is allowed, always within a pre-set limit, in order to facilitate the reporting, the risk to the sponsoring body is low.

The constraints include some compulsory deadlines to be respected within the implementation process of each project in order to benefit from the financial contribution of the sponsoring body (*milestones*). These milestones may include an ultimate external opinion and the delivery of specific documents (e.g. the mid-term report to the sponsoring body, the executive project of restoration, etc.).

The monitoring system detects the progress of the projects within the district. The data collection is entrusted to an external subject appointed by the sponsoring body through a call for tender. This subject works through an on-line platform which is implemented by the working groups directly via Internet. The monitoring outcome results in: periodic reports for the public and confidential technical reports for the Technical Committee and the sponsoring body. These documents are used to assess the performances and adopt the necessary measures if they are negative, such as the reformulation of the management plan and of the budget.

Table 25. Assessment system applied to the Cultural Districts in Lombardy and Veneto (IT): the development factors analysed

Reference study	Districts analysed	Development Factors analysed
Dossena (2013)	Cultural Districts of Fondazione Cariplo	– importance of the district; – size of the district; – age of the district; – governance; – transparency in the construction of the district
Canino *et al.* (2013)	Cultural Districts of Fondazione Cariplo	– institutional coordination; – knowledge accumulation in the social capital of the district; – attractiveness of the territory, cultural offer; – development of professional skills (human capital); – interaction with the other production chains and growth of the local economy.
Ponzini *et al.* (2014)	Cultural Districts of Fondazione Cariplo and Veneto Region (DiCE)	– physical boundaries and spatial configuration; – cultural Market (supply and demand analysis); – management of the district process and involvement of local actors; – funding and monitoring mechanisms.

Source: elaboration by the author

The assessment system provides information about the project effects in terms of local growth. The main issue is the identification of the most relevant factors of development in the district area and the choice of a method for their study on a medium-long term. Generally, an integrated approach is adopted combining various analytic techniques in a common methodological framework, whose nature is strictly connected to the development factors investigated (Table 25).

5.2.2 The feasibility study

The drafting of the feasibility study is a crucial step within the construction of the district, since it outlines the basic contents of the projects. It is divided into three main steps:

1. a detailed analysis of the context (survey);
2. the development of a *masterplan*;
3. the definition of the governance for the future district.

The context analysis is based on an updated framework of knowledge which includes the resources and competences available in the territory as well as the relationships at local scale. The first step consists of the study of the previous cultural projects and of the major local cultural systems. This to find out specific information about: the consistency of the projects; the territorial resources involved in the projects (natural resources, tangible cultural heritage, landscapes, tourist paths and routes, intangible cultural heritage, etc.); the actors engaged in the projects and their nature (public, private, mixed, non-profit associations); the production chains involved (tourism, food and wine, ICT, etc.).

The instruments used include the sociological survey: an appraisal of competencies and resources and, with the purpose of defining the district boundaries, the cluster analysis (Francesconi, 2013). The context factor are considered only indirectly, as seen in the Florence and Marche projects. The study of the spatial distribution, as seen in the Lubjiana case study, is rarely performed.

Subsequently, the analysis of the relational networks and inter-organizational structure within the district area, complete the knowledge framework. In this case, the analytic tools are the social network analysis, the "subject-type of relationship" matrix and the "interdependence of sectors" matrix (Francesconi, 2013).

The information thus collected is summarised to contribute to the development of a district project through the SWOT (*Strengths, Weaknesses, Opportunities, Threats*) matrix as to assess:

- the strengths and weaknesses: internal factors which may foster or impair the construction and launch of the district;
- the opportunities and threats: external conditions which may facilitate or limit the construction and launch of the district.

The results of the SWOT matrix are processed through a positioning matrix in order to formulate the district *masterplan* and its macro-goals (Francesconi, 2013).

The *masterplan* is gradually deepened until it includes: the general and specific goals of the district, the corresponding projects (named "actions"), the stakeholders involved as well as the monitoring and assessment indicators for the whole process. To coordinate the various steps, the GOPP (*Goal Oriented Project Planning*) method is adopted. A project cycle management technique based on workshops and meetings with local actors and intended to progressively implement the following documents (Amari, 2006):

1. "the problem tree diagram": it shows the critical issues within the district masterplan according to the cause-effect relationships arisen during the meetings with key actors and stakeholders.
2. "the goal tree diagram": it is a transposition in "positive" of the previous tree diagram; each current negative situation is turned into a future positive condition, i.e. a design goal leads to specific objectives, the required activities and the expected results;
3. consistency matrix: it is a binary matrix for pair comparisons which assesses the effect of each general goal on the others (positive, negative, or neutral effect) so as to avoid any internal contradiction in the *masterplan*;
4. Logical framework: it is a matrix which includes all the basic elements of the masterplan (targets, results, activities, implementing parties) as well as all the monitoring criteria (indicators, data source, results and expected externalities).

The identification of the district projects and of those responsible for their implementation, is followed by:

- the choice of the local companies in compliance with the profiles described within the logical framework;
- the selection of the most suitable management structure from a legal and administrative point of view;

– the identification of the negotiated planning instruments for the implementation of the projects in the *masterplan*, including the supply of cultural services[17].

The drafting of the feasibility study culminates with the outline of the business plan as well as with the development of the monitoring and control system which assesses the quality and transparency of implementation processes starting from the analysis and design methods chosen by the district, the availability of online information, the presence of a communication plan and performance indicators.

5.3 Conclusions

The eco-systemic approach of the studies on industrial organisation has allowed a better understanding of the creative spatial policies in creative districts.

The analysis of the instruments, procedures and methods used for the construction of the advanced cultural districts confirms the same trend, especially in relation to the development factors investigated. The spatial, economic and social dimensions of the territory are always mentioned in line with the definition of "creative ecosystem" provided by cognitive studies[18]. Although the spatial dimension still plays a minor role, the *Creative Cities* project, for example, is the only one which has developed a methodology for the spatial representation of the creative phenomenon by giving proper weight to the geographic data. Another aspect relatively neglected is the communication and, in particular, its differentiation with respect to internal and external targets, as stressed by place branding theories (Ashworth and Kavaratzis, 2009; Rivas, 2011, 2012). However, these circumstances become clear as soon as the cultural district is regarded as the ultimate evolution of the Marshallian industrial district resulting from the economic studies on innovation within the organization of the production process.

17 For the legal and regulatory aspects of the cultural heritage management in Italy see Camerlengo (2013), Bocci e Ferrari (2013). See also: MiBAC - Direzione Generale per i Beni Archeologici, ed. (2005). *Progetto di definizione di un modello per la realizzazione dei Piani di Gestione dei siti UNESCO*. Roma: MiBAC - Ernst & Young, pp. 189-202 [online] Available on: http://www.unesco.beniculturali.it/Roma [Last access: 24 October 2014.]

18 Although these dimensions may greatly vary, even when the subject of the investigation is the same, as in the Lombard case.

Creative Spatial Policies for Cultural Heritage and Landscape

A Shared Methodology

The analysis of the development of creative spatial policies at international, national and regional scale has highlighted the correlations with the cultural heritage and landscape.

In Italy, the sporadic initiatives of the government are in contrast to the dynamism of cities and regions. The formers are involved in the UNESCO and European programming, while the seconds are building several advanced cultural districts, i.e. complex relational networks acting within different economic contexts based on a multi-sector approach. The creation of advanced districts requires a multidisciplinary methodology which includes analytic tools belonging to social sciences, economics and spatial planning. Furthermore, on the operational level, district policies as well as spatial planning cope with the techniques for the project cycle management competing for the EU and other international funds. In addition, the district masterplans and the urban regeneration strategies[1] are both characterized by knowledge frameworks which consider several development factors showing an eco-systemic approach to planning.

This foresees a convergence between district policies and urban recovery opening new interesting perspectives on the inclusion of the advanced cultural districts within urban planning and the production of new tools for cultural heritage management based on the creative spatial policies.

In the following paragraph we suggests a methodology for the construction of advanced cultural districts whose masterplan coincide with an *integrated landscape project* framed in the landscape planning[2] by regions[3] (Figure 16).

---

1 See Colavitti, A.M. and Usai, A. (2013). Le politiche urbane in Italia e in Europa: dalle sperimentazioni dell'ultimo ventennio alla programmazione 2007-2013, in Colavitti, A. M., *Governance del territorio, beni culturali, piano urbanistico: buone pratiche per la valorizzazione e la competitività*, Firenze: Alinea, pp. 48-75.

2 Here reference is made to landscape planning as described by the European Landscape Convention: "*'Landscape planning' means strong forward-looking action to enhance, restore or create landscapes*" (Council of Europe, 2000, art.1 letter e).

3 Although the methodology is unambiguous, it can be applied in a simplified (preliminary survey, pre-feasibility study), complete (feasibility study) or further detailed version (development of complex programs).

The methodology proposed consists of two basic steps: the assessment of the strategic and organizational positioning and the definition of the *integrated landscape project.* The strategic and organizational positioning helps to delineate the "niche" to be exploited by the potential district in the target market and in the local institutional *milieu* in relation to its territorial capital and reputation.

The main target of the positioning assessment is to determine the future position of the district in the market and, at the same time, the strategy required to reach it (favourable elements, obstacles, organizational decisions, image strategies)[4].

Figure 16. *Procedural steps for the construction of advanced cultural districts starting from an* integrated landscape project *(masterplan)*

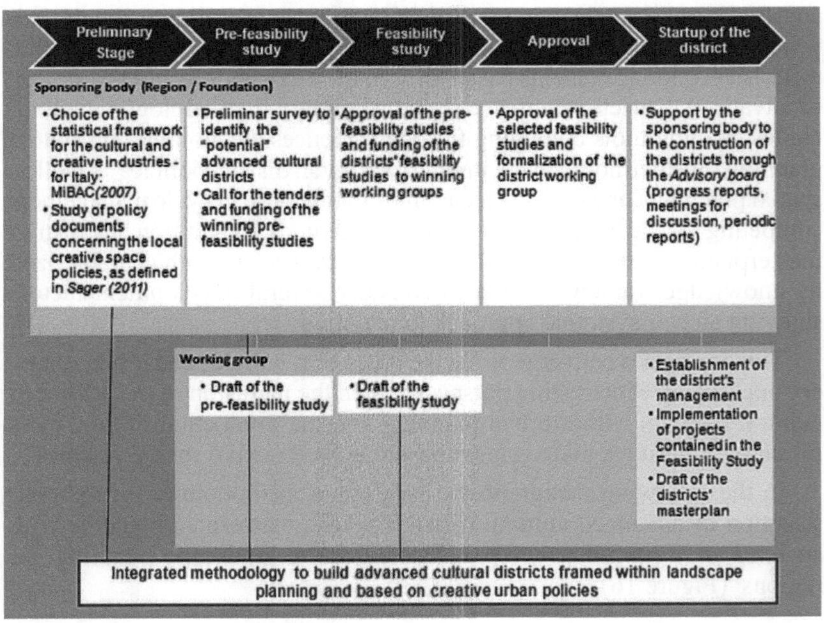

Source: elaboration by the author

4 Since it is connected to the perceptive data (*the district image*), the assessment of the strategic and organizational positioning must take into account the needs of the reference community when it comes to define the basic strategies as well as the intervention policies and actions for the future district (Francesconi, 2013, p. 21).

Within the methodology proposed, the tangible and intangible resources of the territorial capital are analysed according to the techniques developed in the programming of creative districts and in the creative spatial policies (Camagni, 2008). Conversely, the internal and external factors which may foster or affect the construction and launch of the district are examined through a SWOT analysis and a positioning matrix (Dossena and Cioccarelli, 2013).

The second key step, the definition of the *integrated landscape project*, relies on: the identification of the district masterplan together with the structures and mechanisms required for its implementation. The draft of the masterplan is based on the GOPP[5] technique while the definition of the district governance is based on the legal forms outlined by the national laws according to the risks of the masterplan's projects and to the characteristics of the working group.

The following paragraphs deepen the position assessment as well as the construction of the *integrated landscape project* through the description of the steps, instruments and indicators used at this purpose. The final paragraphs illustrate the potential applications of the methodology we propose.

6.1 Assessment of the strategic positioning

6.1.1 Resource analysis

The analysis of the territorial capital of the (future) district relies on the distinction between tangible and intangible resources performed by the Italian regions within the district policies. The **tangible resources** are divided into:

1. Territorial resources: real estate properties belonging to the cultural heritage or containing movable cultural heritage (collections, galleries, archives, museums, etc.)[6];
2. Production resources: industrial areas, existing cultural and industrial districts, local employment system in the cultural and creative sector[7];

---

5 The development of the masterplan starts from the targets, directives and design aspects contained in the regional landscape plan for the territory of the future district.

6 In Italy, reference is made to the definitions of cultural and landscape heritage provided in Legislative Decree n.42/2004 and subsequent amendments (*Cultural Heritage and Landscape Code*), but the investigation is limited to the cultural and landscape heritage characterised by a significant spatial dimension (starting from the architectural to the urban and regional scale).

7 The creative and cultural sectors adopt the definitions established by the local regulatory and statistical frameworks. In Italy, for example, the classification is that provided in the MiBAC White Paper on Creativity (2009).

3. Human resources: actors involved in the local creative spatial policies[8], divided into public, private and mixed subjects (public-private partnerships, jointly controlled companies, etc.), as well as *non-profit associations.*

The analysis of the **territorial resources** draws on the following instruments:

- Database of cultural, landscape and production assets divided by: location; nature (public, private or mixed); type; surface covered; economic and statistical data (funds received, money invested, projects carried out, turnover, number of employees, etc.)
- Geo-referenced cartography (GIS) relating to the resources and to all the geographic factors involved in the landscape planning[9].

The next step consists in the investigation of the **production resources.** Firstly a statistical analysis of the local creative and cultural sector[10] is taken. It has to highlight: the production volume, the main production sectors and production chains; the competitive positioning of the district area with respect to the global, European, national and regional market (city ranking and benchmarking). Then, the analysis of the spatial distribution of cultural and creative companies is performed. Finally, the analysis of **human resources** is drawn through:

- A database of the actors involved in the ongoing planning activities (name, number of projects, financing body, territorial relevance of the project, etc.);
- "Subject-type of relation" matrix, i.e. a summary which links each district actor to the legal, administrative and contractual instruments used to start joint projects with other public or private actors.

The **intangible resources** are particularly difficult to define due to their connection to the knowledge, beliefs and values typical of the local production system as a relational network. The methodology described herein focuses on the skills of governance required to the managers of local cultural network (Table 26). Their quantification inevitably relies on the assessment of skills and resources.

---

8 Defined through a preliminary survey and classified as suggested by Sager (2011).

9 In Italy the techniques and indicators developed within the reference regional landscape plan should be used.

10 Again according to the MiBAC (2009) model.

The actors who were asked to fill in the questionnaires, selected through benchmarking statistical analysis, social network analysis or, in the absence of any preliminary analysis, through the snowball sampling system, had the unique opportunity to testimony the cultural creative industry.

In conclusion, the communication skills of the *network managers* must be considered in order to determine, deliver and promote the reputation of the territory. They can be assessed through an accurate study of the communication strategies adopted by the local cultural organizations[11].

Table 26. Basic competences required to a *network manager* operating in the cultural and creative sector

Skills relating to the management of inter-institutional projects	Coordination of various actors engaged in the draft of the projects. This thanks to the project cycle management techniques and to synergies between the local cultural networks.	
Skills relating to the establishment and management of networks and of "networks of networks" (*network governance*)	*Social innovation*	Ability to progressively build a network connecting actors and stakeholders and fostering the involvement of the economic and social tissue.
	Promotion and brokering services	Ability to establish and control the relationships in order to achieve a balance between problems, needs, opportunities and possible solutions expressed by the actors.
	Benchmarking	Promotion and management of the relationships within other cultural and creative systems inside and outside the district (relations with other networks).
	Benchlearning	Integration between the techniques for the performance improvement and those for the organizational learning.
	Change management	Ability to stimulate and support the changes in the redefinition of the "game rules" at the institutional and governance scale.
	Programming and control	*Accountability* and social balance reporting.

Source: elaboration by the author on Francesconi (2013)

11 Here reference can be made to the feasibility study for the construction of an advanced cultural district in the Province of Ascoli Piceno (IT) and especially to the section entitled *La comunicazione e il Territorio di Ascoli Piceno* (Sacco, 2007, pp. 48-59).

6.1.2 Relational analysis

The advanced cultural district is a particular form of advanced industrial district characterised by a variable organisation, systemic interdependences, several integration options (vertical, horizontal and transversal) between the cultural and production chain, thus leading to the establishment of "hybrid" districts relating to technological innovation, tourism and food industry. It includes various networks whose nodes correspond to public, private or mixed cultural institutions and companies: from cultural networks to local cultural systems, from technological clusters to traditional cultural districts up to meta-districts.

The assessment of the positioning of a future district has necessarily to consider the relationships existing among the production resources, the cultural heritage chain, the other production chains and the actors involved in the local creative spatial policies. This is possible making reference to the instruments shown in Table 27.

Table 27. Analysis of the inter-organizational relations

	Investigation instruments	**Indicators**	**References**
Relations among the production resources involved	Context analysis	Environmental and management indicators	*Municipality of Genoa (2013)*
		Design indicators	*City Form Lab (2014)*
		Indicators for cultural districts	*Carta (2004)*
		Indicators relating to the standards for public services in urban planning tools based on zoning	*Literature on land-use zooning*
	Group analysis (inter-sector relationships within the cultural chain)	Statistical indicators for clustering	*Maggino (2005)*
		Indicators of spatial proximity or distances	*Maggino (2005)*
	Social network analysis (SNA) (limited to companies)	Network density	*Lazzeretti (2012)*
		Centrality degree Betweenness centrality degree	*Lazzeretti (2012)*
		Centralisation degree Betweenness centralisation degree	*Lazzeretti (2012)*
	Subject-type of relation matrix (limited to companies)	/	*Sacco (2007)*

	Investigation instruments	Indicators	References
	Group analysis (relations among the cultural heritage chain and the other local chains)	Statistical indicators for clustering	*Maggino (2005)*
		Indicators of spatial proximity or distances	*Maggino (2005)*
	"Interdependence of sectors" matrix	/	*Francesconi (2013)*
Relations among the actors involved	*Social network Analysis* (relations among the subjects involved in the same public project)	Network density Centrality degree Betweenness centrality degree	*Lazzeretti (2012)*
		Centralisation degree Betweenness centralisation degree	*Lazzeretti (2012)*
	Subject-type of relation matrix (relations among the subjects involved in the same public project)	/	*Sacco (2007)*

Source: elaboration by the author

6.1.3 Evaluation of the strategic-organisational positioning

The information collected through the appraisal of the resource and relationships are summarised and used for the construction of the SWOT matrix (*Strengths, Weaknesses, Opportunities, Threats*) (Figure 17).

The SWOT matrix helps outline the current strategic-organizational positioning of the potential district, however, the information provided must be further processed to formulate the district's masterplan and define its macro-goals. The instrument used in this step is the positioning matrix (Figure 18).

Figure 17. *The SWOT matrix*

	Internal environment	External environment
Potentials	Strengths	Opportunities
Risks	Weaknesses	Risks

Source: Francesconi (2013)

Figure 18. *Strategic-organisational positioning matrix*

		Strategic relevance	
		High	*Low*
Relative strength	*High*	Crucial strengths	Redundant strengths
	Low	Critical weaknesses	Non-relevance area

Source: Francesconi (2013)

6.2 The *integrated landscape project*

6.2.1 Definition of the project "Masterplan"

The district's masterplan relies on a first strategic document drawn up according to the positioning matrix and then progressively deepened through the project cycle management techniques, like the GOPP (*Goal Oriented Project Planning*) methodology[12]. It is structured on the following phases[13]:

1. Definition of the Logical Framework: identification of the masterplan contents (general goals, specific goals, targets); definition of the actions as well as of the indicators of impact and result (Table 28);
2. Construction of internal consistency matrix for the logical framework goals: it assesses the consistency of the general and specific goals with respect to the problems identified, of the results and activities in relation to the specific objectives, as well as of the indicators with respect to the results to be assessed;

12 For further information on the GOPP approach, see Amari (2006, pp. 130-143), Bussi (2001, 2002) and European Commission (2001).

13 Some of them can result from the environmental report issued within the Strategic Environmental Assessment of landscape planning tools and from the feasibility studies of the existing districts.

3. construction of the external consistency matrix for the targets included in the logical framework: analysis of the consistency of the goals indicated in the logical framework with respect to the specific objectives reported in the local urban planning tools as well as in the local creative spatial policies which may affect the district[14];
4. identification of the local stakeholders to be involved in the construction of the district and of their competences and resources: their skills can be detected through an assessment of human resources as well as of the relationships among the actors involved in the local creative spatial policies.

The consistency matrix origins in the Strategic Environmental Assessment (SEA). It

The assessment matrix reports the targets/actions from the masterplan in its lines and the specific goals/actions of the local planning tools/ creative spatial policies in its columns. The alignment of a masterplan target/action with the goal/action of an external planning tool leads to a "consistency" grade, vice versa, an "inconsistency" grade is given when the masterplan choices affect the implementation of the goals of the external planning tool. When the masterplan decisions do not affect those contained in the external planning tool, an "indifference" grade is assigned.

The consistency, inconsistency or indifference grades can be expressed through graduated scores according to rigid ordinal rating scales (for ex. a score from 1 to 3; 1 if the operating target is irrelevant to the specific objective, 2 if it is relevant, if 3 it is decisive), or by means of multi-criteria analyses and interference matrices (more complex techniques which reduce the subjectivity related aspects). At the end of the analysis of consistency, the goals of the masterplan and actions which are inconsistent with the external planning tools are subjected to a process of harmonization.

[14] For the analysis of the external consistency within the environmental impacts assessment see Conzimu and Zoppi (2008), ISPRA and MATTM (2012), Torre *et al.* (2011).

Table 28. Logical Framework

	Intervention purpose	Indicators	Data sources	Assumptions
General objectives	Long-term social and/or economic benefits for the whole society provided by the project.	Non-subjective measurable parameters which record the achievement of a goal (or of a result). They include the following elements: – a variable – a *target-group* – an observation period – a reference value.	Sources of the data used to develop the indicators	Risk factors for the project: external conditions, whether positive or negative, which must occur to allow the achievement of targets and results by the project
Target (specific goal)	Any aspect and condition of the beneficiaries' life which will benefit from the use of the services provided by the project.	Indicators	Data sources	Risk factors for the project
Results (continuation)	Knowledge and skills that the beneficiaries will acquire thanks to the project activities (including the services offered and excluding the infrastructures)	Indicators	Data sources	Risk factors for the project
Activities	Aspects or conditions of stakeholders' life that will improve as a result of the project.	Resources to be invested	Data sources	Risk factors for the project
Basic conditions Regulatory framework promoting the interactions between public and private actors and the appointment of cultural services to external individuals.				

Source: elaboration by the author on Amari (2006) and Bussi (2002)

6.2.2 Identification of the governance model

Once the project actions and the responsible of the project implementation are identified, it is possible to define the governance of the future district, i.e. the management structure most suitable under the legal and administrative points of view.

Every decision and agreement is usually preceded by a letter of intents, i.e. a series of documents attesting the commitment of the signatory parties in the future district and in its *integrated landscape project* (masterplan). The letter of intents is used both to strengthen the working group and to formalize the links between the working group and the other actors involved in the *integrated landscape project* (public administrations, trade associations, companies, etc.).

The letter of intents is the starting point for the definition of political and technical agreements between the stakeholders.

The technical agreements determine the nature of the ties connecting the various actors engaged in the construction and launch of the district and they allow for the shaping of the management structure of the future district (Amari, 2006, pp. 140-141). The choice is limited to the management models approved by the national and regional regulatory framework. For instance, the Figure 19 shows the management models for heritage valorization recognized by law in Italy. The potential and limits of each method are illustrated through a SWOT matrix in the guidelines provided by the Ministry of Culture for the drafting of the management plans in UNESCO sites (MiBAC, 2009).

After the definition of the district governance, there is the identification of planning tools necessary to implement the measures included in the *integrated landscape project* (Table 30).

Figure 19. *Valorisation of cultural heritage in Italy: management forms according to the national laws*

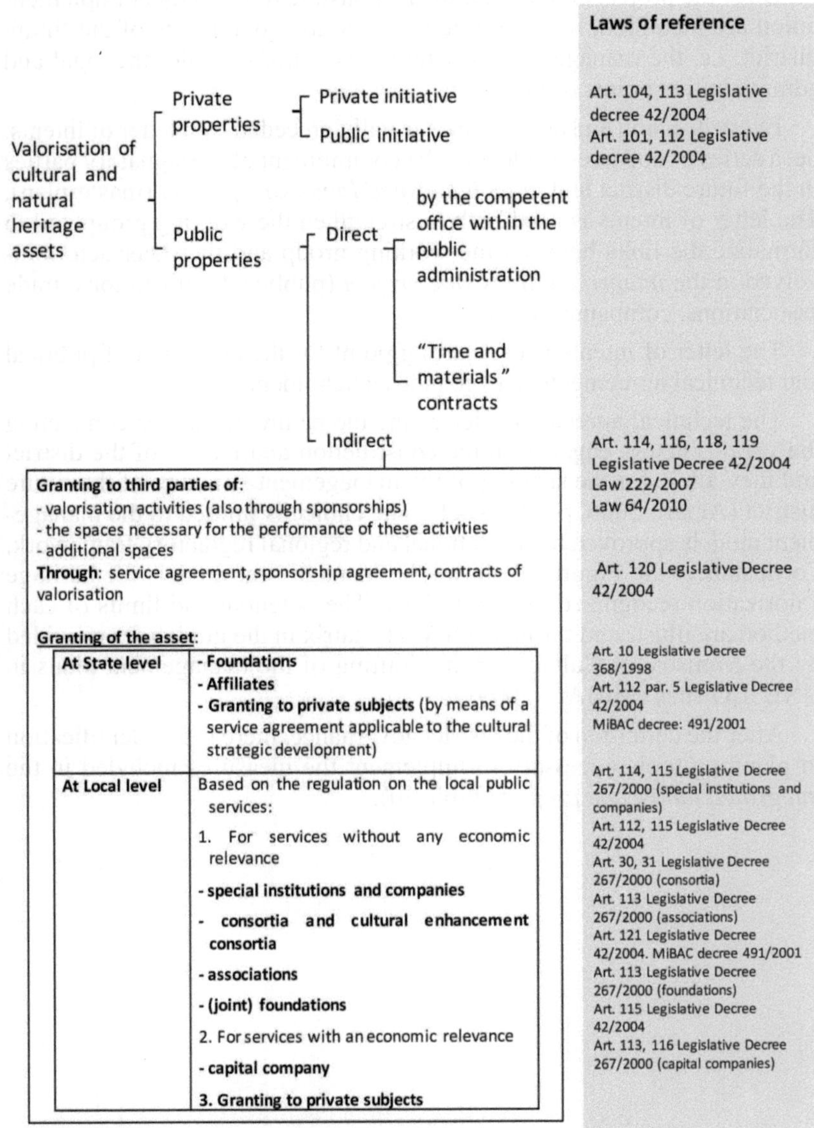

Source: elaboration by the author on Camerlengo (2013), Bocci and Ferrari (2013)

Table 29. SWOT matrix for the assessment of the legal forms identified for the management of the future district

Legal form	Advantages	Costs	Strengths	Weaknesses	General adaptability to the site
Consortia	Easy establishment of the body, since it is sufficient to sign an agreement. Consortia can also promote business activities	Limited, since the consortium has no minimum social capital and no accounting and fiscal obligations are required.	The structure is flexible, since the civil code grants private bodies the maximum freedom. Both public and private entities can participate.	The structure is less dynamic than capital companies, if the site requires a business management	Good, thanks to the flexibility of the consortium and the possibility to involve public and/or private subjects. Very good, when there are numerous competent bodies.
Conventions of local bodies	There is no need to establish a new legal entity	Limited, since the existing public structures are used.	Low establishment costs, good coordination ability, provided that the local bodies are few.	The private entities cannot participate. The structure is poorly dynamic. The failure to establish a new legal entity may affect the access to any financing (which could conversely be assigned to single entities). It cannot be applied when State entities are involved.	Good, if the site responsibility rests with few local entities. Moreover, only the local entities must comply with this responsibility.
Non-profit recognised associations	No minimum social capital	Low management costs and simplification of the civil and tax requirements	Both public and private entities may adhere to the association. Tax advantages.	It is not suitable for any operational management and/or the achievement of economic targets.	Good, since the association rules out the possibility to carry out business activities and its statute provides for the protection of the site. Good especially for coordination activities.

Legal form	Advantages	Costs	Strengths	Weaknesses	General adaptability to the site
Capital companies	Independent legal personality and imputation of interest.	High for joint-stock companies and lower for limited liability companies. Obligation to comply with the civil and tax requirements relating to company laws.	Capital companies are characterised by a dynamic and active management of the corporate purposes, thus increasing the available financial resources used to produce goods and services. Possible financial contributions by private subjects.	They can poorly adapt to the peculiarities of the cultural heritage to be protected and their non-profit nature. The corporate purposes do not coincide with those of each member, which therefore cannot act unilaterally through the exercise of authoritative or discretionary powers. In case of multiple competent bodies (both local and state), the implementation of this instrument may be difficult based on the applicable regulation and specific provisions should be required.	Good for the promotion and/ or development of business activities. Good for the management of the asset(s) with a business approach, also enhancing and promoting the site in terms of tourism.
Foundations	Ability of the local bodies to directly provide cultural services	Low management costs	Low accounting and tax requirements	Legal instrument not suitable for UNESCO sites. The assets are intended exclusively for the accomplishment of the targets.	Low, since the structure of foundations does not allow for the aggregation of various subjects. It should be applied to the assignment of cultural services (libraries and newspaper libraries), i.e. in case of indirect management of the cultural heritage as per art. 115 of the Cultural Heritage Code.

Legal form	Advantages	Costs	Strengths	Weaknesses	General adaptability to the site
Special local company	A subject directly appointed by the local bodies for the management of the services	High due to the fact that the local company is managed like a capital company.	Particularly suitable for UNESCO sites, the responsibility resting with a single local body.	Other local and/or national bodies cannot aggregate and private subjects are excluded.	Low, considering that the responsibility for UNESCO sites are unlikely to rest with just one body. Low due to the inability of different subjects to aggregate.
Consortium entities	Independent legal personality and imputation of interest.	High for joint-stock consortium entities and lower for limited liability consortium entities. Obligation to comply with the civil requirements.	Consortium capital entities are characterised by a dynamic and active management of the corporate purposes. Possible financial contributions by private subjects.	Legal instrument poorly suitable for the management of UNESCO sites, beyond the assumptions provided in art. 113 bis of the TUEL relating to the subjects who carry out commercial activities.	Low, since the structure of the consortium entities must be made up of professional traders. It may be applied to non-profit activities aiming at the protection of the site.

Source: MiBAC (2005)

Within the traditional feasibility studies, the district's masterplan results in architectural and environmental interventions aimed at the recovery, maintenance and accessibility of the cultural and natural sites as well as the services for the public and the coordination of the parties involved (central government, local bodies, authorities, managers, tour operators, etc.). However, in the methodology we propose, the masterplan is an *integrated landscape project* which is thought to join the landscape planning by the regions, the regional policies for the cultural districts and the local creative spatial policies relating to cultural heritage. The purpose is to define, at least at the operational scale[15], a consistent framework for the spatial, economic and social changes in the district area. Therefore, the masterplan "based on projects" of the traditional cultural district is here replaced by a new one "based on programs". It relies on the national and European programs to cope with the cultural heritage issues with a holistic

[15] Considering that, from a theoretical and administrative point of view, urban planning and the process of creating districts are two independent processes.

approach and increase the funds available[16]. Within the European programs, reference is made to: INTERREG and URBAN for the 2000-2006 programming; the Joint European Support for Sustainable Investment in City Areas (JESSICA) financial instrument for the 2007-2013 programming; the Integrated Territorial Investments (ITI) for the 2014-2020[17] programming (European Commission, 2013, 2014; Usai, 2013).

In case of Italy there are two groups of tools: the first one includes the local development programmes dedicated to urban regeneration, while the second one consists of plans on urban renewal and urban regeneration (Cremaschi, 2001; Cremaschi, 2002; Governa, 2004; Saccomani, 2004; Forgione, 2008; Tomei, 2013). Finally, the programs for cultural heritage valorisation introduced by the L.D.42/2004[18] (Table 30).

---

16 The integration of the advanced cultural district into the landscape planning and the implementation of the integrated landscape project (masterplan) by means of complex programmes may help the district benefit from the resources arising from the Community programming (HORIZON 2020, COSME; NOP and ROP), the landscape planning, the national urban policies (for ex. the Italian National Plan for the Cities), the ministerial initiatives (in Italy, the MIUR e MiBACT calls for tenders) and the regional economic programming (regional development programmes).

17 Thanks to the JESSICA financing as well as to the NOP and ROP structural funds, since 2009 the Italian regions and cities have financed several medium- and long-term integrated urban recovery plans whose identification for study purposes is affected by their classifications and contents. In Tuscany, for example, Integrated Plans for Sustainable Urban Development (PIUSS) have been financed and adopted to turn abandoned or suburban city areas into new urban centres able to drive extra-urban factors – at least, where possible, thanks to special conditions such as the presence of monuments or relevant facilities (Pingitore and Rignanese, 2009). Conversely, in Sardinia, thanks to the 2007-2013 ERDF ROP funds, the Integrated Urban Development Plans (PISU) and the Sustainable Energy Action Plans (PAES) have been introduced – i.e. instruments for the implementation of strategic municipal and inter-municipal plans financed through the 2000-2006 ROP (RAS - CRP, 2012).

The Integrated Territorial Investments (ITI) of the 2014-2020 programming are similar to the JESSICA financial instruments, but allow for the financing of integrated territorial projects drawing on NOP and ROP resources or on different European programmes. This has further complicated the study of the integrated urban and environmental recovery plans financed by the EU funds. For further information on this topic, see: Regulation (EU) n.1299/2013 art.10; Regulation (EU) n.1301/2013 art.7; Regulation (EU) n.1303/2013 art.32; Regulation (EU) n.1304 / 2013 art.12.

18 In Italy, for example, within the National Plan for the City, some urban enhancement agreements have been introduced (e.g. in the city of Ancona, Bari, Naples and Pavia). The Code of cultural heritage and landscape, however, has introduced specific enhancement programmes as well as Strategic plans for cultural development (Legislative Decree 42/2004 article 112). The latter have been adopted in the disposessions of the *federalismo demaniale cuturale* (see paragraph 3.3.2). (Colavitti and Usai, 2013a, 2013b, 2014).

Table 30. The tools of negotiated planning to implement the interventions included in an *integrated landscape project* in Italy

EU programmes (NOPs and ROPs)	– **Sustainable development plans and Strategic plans** under the 2007-2013 EU programming (e.g. PIUSS in Tuscany and PISU in Sardinia) – Integrated territorial interventions (ITI) under the 2014-2020 EU programming – Local Agenda 21
Government initiatives	– Territorial and urban strategies relating to **Big Events** (e.g. urban strategies and projects for EXPO 2015 in Milan) – Contract of valorisation within the **National Plan for City**
Plans for urban recovery and urban regeneration established by national laws	**Urban regeneration programmes through local development:** – territorial plans – local and programming agreements – local employment plans
	Urban regeneration complex or integrated programmes – Plans of Urban recovery – PRU (1994) – Urban renewal plans – PRIU (1994) – Urban renewal and local sustainable development plans – PRUSST (1998)
	Integrated instruments for the cultural heritage (L.D.42/2004) – Programs of valorisation (also for the *federalismo demaniale culturale)* – Strategic plans of cultural development

Source: elaboration by the author

An *integrated landscape project* requires also the development of a business and communication plan.

The business plan turns the organizational, technical, production and commercial information contained in the feasibility study into the following documents: the balance sheet of the district; an account of the financial receivables and payables of the district; financial statement; initial budget; expected costs for the sites' recovery; synthetic report on equities, incomes and finances connected to the tangible and intangible assets, to the interventions as well as to project management (Amari, 2006; Sotti, 2013).

The communication plan defines the promotion initiatives and instruments used to build an effective reputation. A competitive identity which helps the district to find the proper position in the target market (Anholt, 2011). The plan may be either drawn up directly by the working group or assigned to a firm specialising in place branding[19]. During the construction of the competitive identity of the future district, it is necessary to distinguish between the following forms of communication (Ashworth and Kavaratzis, 2009; Kavaratzis, 2004; Rivas, 2012):

1. Controlled communication: messages directly delivered by the working group or the appointed firms. It is divided into:
   - Primary communication: it refers to the communicative effects resulting from the decisions made by the district team during the works in terms of: landscape (public spaces and urban planning, architectural features, built environment); infrastructure (accessibility); organisation (management of the relations between stakeholders and their involvement in the decision-making process); behaviour of the working group (services, and events for inhabitants and visitors, efforts made by local authorities to involve the actors in the organization of these initiatives).
   - secondary communication: it refers to the official communication of the district backed up by the working group through advertising, public relations, design, etc.
2. uncontrolled or tertiary communication: uncontrolled messages delivered by the inhabitants and the competing territories by "word of mouth", also influenced by the primary and secondary communication.

The construction of a competitive identity for an advanced cultural district mainly relies on controlled communication in the hope that it may lead to a positive tertiary communication. Moreover, since it is hard for the working group to directly affect the urban and infrastructure planning, the main focus is on the organizational and behavioural choices involved in primary communication, which are particularly relevant for the establishment of external and internal alliances allowing the district to retain its future international reputation even in transition phases (e.g. the change of leadership in the district, a chaotic institutional framework, political changes in local government, etc.) (Bellini *et al.*, 2010; Hankinson, 2010; Kavoura, 2013).

In the choice of the communication strategy for the district the working group should (Figure 20):

---

19 The place branding is a multidisciplinary approach to territorial marketing. It arises from urban and territorial promotion but takes inspiration from marketing, especially from business marketing. Business marketing analyses the relations between the brands and the reputation of the producers as well as between the latter and the local environment. This has increasingly connected marketing assumptions with the peculiarities and needs of the territory (Kavaratzis and Ashworth, 2010; Kavoura, 2013).

- create a specific strategy, i.e. develop a clear idea of the district identity and its current and future positioning based on the territory image resulting from local products and production systems, the others' opinion about the district and the self-conception of the district (Anholt, 2011);
- focusing on primary communication, they should establish alliances with local and *supra*-local actors through formal agreements or joint activities, with the purpose of ensuring the contents and the symbolic actions necessary for a partnership building strategy (Hankinson, 2010).

As anticipated, the working group can assign these tasks to external firms (*place branding organisations* – PBOs) which are responsible for the primary and secondary communication in the district.

Figure 20. *The sources of communication for the construction of an advanced cultural district*

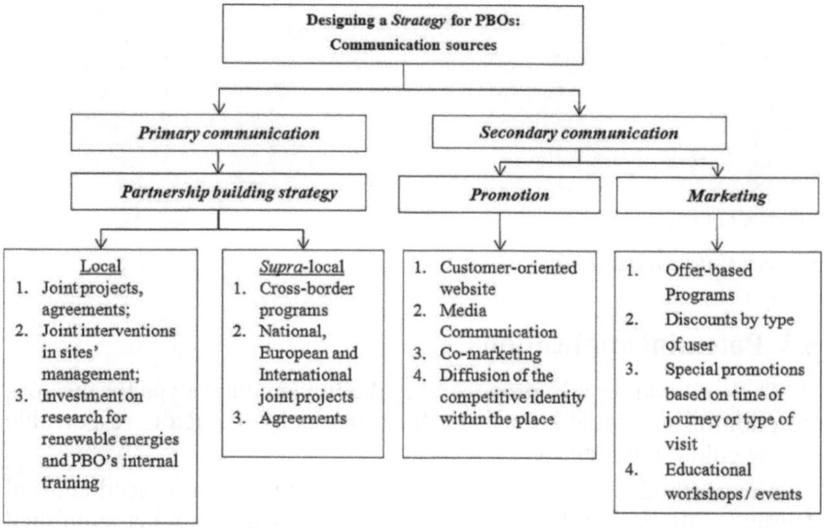

Source: Colavitti and Usai (2014a)

The Figure 21 shows the responsibilities of these firms in the construction of an advanced cultural district according to the methodology we propose.

Once the communication plan has been defined, the *integrated landscape project* is defined. The last step is the development of the monitoring and control system as to guarantee the quality and transparency of the start-up phase. This happens through: on-line information, communications tools, indicators of *performance* (taken from the indicators selected for the logical framework of the feasibility study).

Figure 21. *The strategies adopted by place branding organisations for the construction of an advanced cultural district*

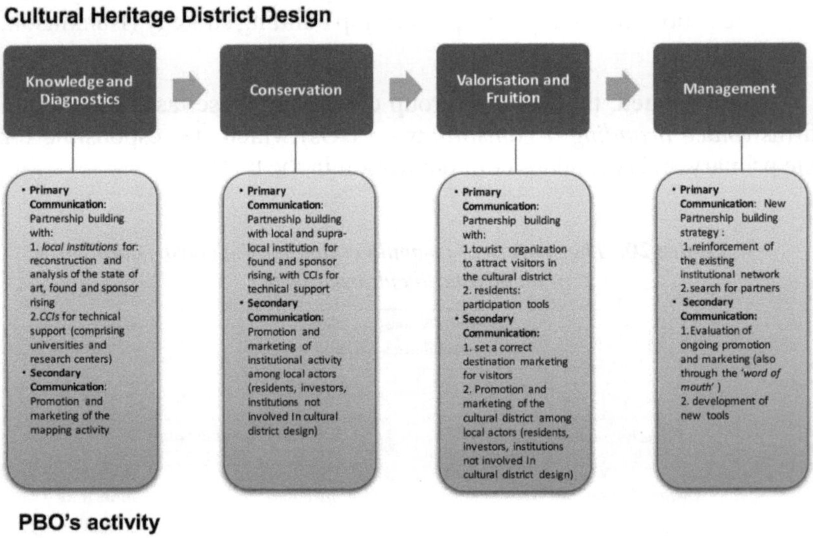

Source: Colavitti and Usai (2014a)

6.3 Potential applications

The methodology we propose fixes the fundamental steps for the construction of integrated knowledge frameworks by the actors responsible for the cultural heritage and landscape.

This methodology is conceived for the creation of advanced cultural districts at the regional scale but it can be also applied, either simplified or extended, at the urban scale. Even cities, in fact, increasingly recur to integrated plans for the recovery and regeneration of the historic urban landscape. Their implementation relies on wide masterplans, similar to the *integrated landscape project* illustrated.

New applications are expected for this methodology, for example in the updating of the existing plans or programs on cultural heritage and landscapes that can be refinanced under the 2014-2020 EU programming or under the government initiatives.

Furthermore, this methodology can also be used for the construction of other kinds of district linked to cultural heritage, such as the advanced tourist districts as well as the technological districts and clusters for heritage conservation and valorisation.

A First Application

A Shared Methodology for the Management of Cultural Heritage in Sardinia Region (IT)

After clarifying the contents and the possible uses, now we try to highlight the potentials and the limits of the methodology illustrated in chapter 6.

In the following paragraphs we compare the analytic techniques of the methodology to those currently used in the draft of integrated landscapes projects. In particular, we focus on the landscape projects carried out by Italian regions which do not have advanced cultural districts but which are provided with advanced planning tools for cultural heritage and landscape.

In Italy the Sardinia region (IT) has committed itself to the creation of cultural districts financing, between 2004 and 2008, a feasibility study and approving a specific law on the matter. This has given the start to a reform of the regional cultural policies. However, the election of a new regional council in 2009 overrode the start of the cultural districts, which remained on paper.

During this time, the Region updated the planning system with the new rules introduced by the Code on Cultural Heritage and Landscape (L.D.42/2004), adopting in 2006 the Regional Landscape Plan, i.e. the regional tool for landscape planning – as defined in ELC – which defines and rules the contents of all the urban development plans. In addition, the partial isolation due to the insularity and to the statutory autonomy, makes Sardinia a kind of "option zero" scenario as regards the creation of advanced cultural districts and, consequently, the ideal ground to test the methodology we propose.

Starting from an interpretation of the urban plan as an instrument to coordinate the relationships between the territory, the population and the economic activities, the first part of the chapter analyses the techniques and the instruments used for the analysis of these three dimensions in Sardinia landscape planning. The aim is to clarify whether there is an irremediable gap between urban planning and the public policies for the districts or whether there are some points of contact that anticipate its overcoming.

The second part of the chapter illustrates the techniques used by Sardinia Region in the construction of integrated landscape projects and how they refer to those of our methodology (Chapter 6). The comparative analysis is done through a theoretical framework which puts the two methodologies side by side and analyses their similarities and differences in each step (Table 31). The qualitative judgements used in the framework, which are *present*, *partly present* and *absent*, refer to the presence or not of that particular step or content of the methodology we propose within the methodology adopted by Sardinia Region.

In the final part of the chapter, we discuss the findings of the comparative analysis and the future prospects in research.

Table 31. Framework for the analysis of the integrated landscape projects

Iter for the construction of advanced cultural districts	***Iter* for the draft of the Integrated programs for landscape by Sardinia Region**
A. Preliminary stage	
B. Call for tenders for the selection of pre-feasibility studies	
C. Funding of the feasibility study	
D. Approval of feasibility studies	
E. Start-up (implementation of interventions of the feasibility study)	

Drafting of the feasibility study	**Drafting of the integrated landscape projects (masterplan)**
1. Analysis of the resources	
2. Analysis of the relationships	
3. SWOT matrix for the strategic positioning (*benchmarking*)	
4. Construction of the masterplan (= integrated landscape project) and its macro-goals, general goals, actions, subjects involved	
5. Costruction of the *governance*:	
5.1 analysis of the management forms provided by the current laws;	
5.2 choice of the planning instruments to carry out the actions set in the integrated programme of landscape;	
5.3 final choice of the management form	
5.4 economic-financial plan, communication plan, definition of the monitoring system	

Source: elaboration of the author

7.1 The plan as an instrument for the regulation of the relationships between territory, population and economic activities

In urban planning, the inclination towards an integrated approach to solve territorial issues is present since the end of the 19th century with the Regionalist School, represented by scholars such as F. Le Play, F. Ratzel, P. Kropotkin, E. Reclus, P. Vidal de la Blache, P. Geddes. The regionalists used the term "region" not with the meaning as an administrative unit but, instead, as a geographical unit: a socio-economic-territorial unit typified by constructive and destructive dynamics that ensure its equilibrium in time and space (Mercer, 1997; Scimeni, 1979).

Notably, Patrick Geddes suggested a synoptic approach to regional planning: it consisted in replacing the monodisciplinary approach with a comprehensive and concise knowledge, able to return the complexity and vitality of the real life. This is also valid for the city and for its "spirit". Geddes defined it as the intricate, ancient and always new fabric built over time by the intertwined relations among place, work and people (Gaeta *et al.*, 2013, pp. 156-158). This concept strongly influenced the following studies because of its connection with the themes of history[1] and participation[2].

In the 1933 book *Town and Country Planning* Patrick Abercrombie presents his contribution to the *survey* by Geddes: twelve items and forty-two elements "of detail" that, according to him, could have been a part of the regional survey (Matless, 1992). Through his work, however, Abercrombie promoted urban planning as an autonomous discipline (and profession)

1 Geddes' historical view was adopted by the historicist and aesthetic studies, even if considering only the aspects related to the safeguard of the ancient city (*conservative surgery*) and to the aesthetic of new developments, as in the works by Camillo Sitte (1889) and Gustavo Giovannoni (1931) (Gaeta *et al.*, 2013, p. 153).

2 The "spirit of the cities" of Geddes, in fact, refers to the perception the inhabitants have of their environment and to its evolution in time and space.
This concept spread widely in the post-war period thanks to Lewis Mumford (1960), a disciple of Geddes. He stressed the survey as the core of Geddes' proposal, promoting a "scientific" regional planning based on the direct observation of the spatial, economic and social phenomena (Mercer, 1997; Scimeni, 1979; Veldpaus *et al.*, 2013). The "spirit of the cities" idea influenced the imageability by Kevin Lynch and the works by Benedict Anderson (1983) on the region as a culturally-defined geographical unit – an "imagined community of practice". Recently, it has influenced the research by Homi Bhabha (2005) on the third space and also the works by Charles Landry (2006) which has defined the civic ethos of the creative city as a sensorial horizon, broad but critical, tied to the individual and collective experience.

within regional[3] planning putting some distance between his work and the Regionalist School (Gaeta *et al.*, 2013, p. 185)[4].

In the post-war period the national economic planning and the urban planning developed in parallel as branches of regional planning[5]. The economic planning provided to urban planning a coherent picture to sort out problems and to define the objectives and contents of the public policies. Vice versa, the urban planning guaranteed a correct implementation of the economic policies checking dimension, quality and localisation of the residential and industrial settlements – both new and old (Scimeni, 1979).

Unfortunately, the eco-systemic approach dead out in facing the various issues still open (the reconstruction after bombings, the increase of housing demand, the unauthorized building, etc.). Economic and urban planning were no longer parts of the broader filed of regional planning, despite they affirmed themselves as autonomous disciplines[6] (Astengo, 1967; Benevolo, 1963; Mercer, 1997).

---

3 *Territorial* when addressed to wide territories, *urban* when focused on the city, *operative* when it refers to quarters or little urban areas (Gaeta *et al.*, p. 188).

4 It is exactly in the *Greater London Plan* that the first signs of the reduction of regional planning to urban planning, seen as land-use planning, with all its criticalities, begin to be seen (Gaeta *et al.*, 2013, p. 224).

5 This phenomenon takes place in Italy too, with the Consortia for industrial areas funded by the *Cassa per il Mezzogiorno – Comitato dei Ministri per il Sud* (Development Fund for Southern Italy – Committee of Ministries for Southern Italy), the Regional Planning Centers and the *development districts* created by the Ministry of Industry and of Commerce (1962) and by the Ministry of Finances (1964-1968). Other interventions must be added to these: the reform of the public administration and of the planning system converge in the 1974 rules for the start-up of the Ordinary Regions, in the creation of Provincial and Regional Coordination Territorial Plans, in the formulation of institutional *performance standards* to become part in the European Economic Community (Colavitti, 2013, Fried, 1968).

6 John Friedmann (1974, 1987) and Andreas Faludi (1973), in outlining a picture of planning from the 1930s to the 1980s, represent a variegated scenario in which they spot, more or less distinctly, the signs of a disciplinary crisis tied to the progressive separation between substantive planning (*planning theory*) and procedural planning (*theory in planning*). Considering planning as a method and scientific instrument to validate the necessary knowledge for action, Friedmann and Faludi reaffirm how theory and practice are both necessary for an effective planning but their paths reach different outcomes. Friedmann (1987) suggests a trans-active planning, based on an action that socially transforms the community, which acknowledges the failure of the public participant's authority in modern society. Conversely, Faludi suggests an integrated and systematic approach to substantive planning through the inception of a new discipline, the science of planning or *planology*, in his conviction that this may have significant consequences on the practice of plan and on disciplinary instruments (Archibugi, 1998; Carta, 1996; Thomas and Healy, 1991).

The region as an elementary unit of any form of planning returned at the centre of the international debate only with the report *Our Common Future* of the World Commission on Environment and Development which introduced the concept of sustainable development (United Nations, 1987, art. 40). Thus, the "sense of place" of Patrick Geddes was re-discovered. The perceptive data were used to update the techniques of representation, preservation and valorisation of the cultural and natural heritage as to facilitate the mapping of regional cultures[7] (Anderson, 1983; Mercer, 1997; Schama, 1995). The ecosystemic approach to planning of the Regionalist School tied to the organic metaphor and to the historical-perceptive data. Finally, it has been officially recognised through the 2000 European Landscape Convention and spread in Europe through the landscape planning (Council of Europe, 2000)[8].

In last year the "technological revolution" is reshaping the economic, social and spatial fabric like the industrial revolution did in the 20th century. This will test the compliance of the ecosystemic approach by landscape planning to the new issues and challenges of regions.

7.2 Planning in Italy[9]

The recent economic crisis has dramatically reduced the spending power of public administrations, forcing them to concentrate the interventions on the most important assets of the cultural heritage and landscape. These are in fact, seen more and more often by public administrators as rare and irreproducible economic goods which allow a region to compete globally (Crivello, 2010; Florida, 2002, 2005).

The Italian cultural heritage and landscape, which are wide and multilayered, could not escape these logics and they have therefore become a central topic in public policies (Amari, 2006; Carta, 2002, 2004, 2007; Sacco *et al.*, 2006; Scandale, 2005).

---

7 Monodisciplinary approaches have proven to be ineffective in representing how the settled communities perceive themselves, how they are perceived by the others or how they think they are perceived by others, also from the point of view of the identity heritage, i.e. the cultural heritage and landscape which the local community identifies with.

8 " *'Landscape' means an area, as perceived by people, whose character is the result of the action and interaction of natural and/or human factors*" (Council of Europe, 2000, art.1 letter a).

9 The following paragraph is a re-elaborated version of the contributions drawn and discussed in three different international conferences: *Regional planning and territorial competitiveness: the role of Identity Heritage. The case of the Sardinian Region* (Colavitti *et al.*, 2013a); *Debating the Landscape. The Role of Landscape Regional Planning Between Local Community Identity Project and Institutional Landscape Idea* (Colavitti *et al.*, 2013b); *Distinctive and Innovative Regions: Urban Resilient Communities and Regional Policies. The Sardinia Case Study* (Colavitti *et al.*, 2014).

After the European Landscape Convention cultural heritage and landscape have become the starting point for new forms of government of the territory characterised by the "right to landscape", generally referred to as landscape planning[10] (Council of Europe, 2000; ESPON, 2013; Gambino, 2009).

In Italy the Convention has been implemented thanks to the Code on Cultural Heritage and Landscape (Legislative Decree 42/2004, hereafter the Code) which has updated the system of safeguard. The Code devolves the competences on heritage valorisation to Regions but it also imposes the adoption by the same of Regional Landscape Plans (RLPs), i.e. plans of regional level overarching the urban planning tools adopted by cities and towns[11], so as to resolve the historical conflict between city planning and safeguard of the cultural heritage, born with the 1942 Town and Country Planning Act[12] (Barbati, 2008; Sciullo, 2008; Cartei; 2008).

The RLPs approved since 2004 show how the Regions have interpreted the prescriptions of the Code according to the needs of their territory, to the statutory situation and to the maturity of the local urban policies. This emerges especially in implementing the art.143 of the Code which gives the Regional Landscape Plans the possibility to safeguard new categories of heritage assets and landscapes in addition to those protected by the national laws. This happens on the base of their identity value according to the local community[13] (Barbati, 2008; Cartei, 2008; Castiglioni, 2009; Gambino, 2003; Ponzini, 2008; Sciullo, 2008).

The RLPs' draft and/or revision is still in progress and their participatory instruments have not been fully tested in the implementation stage. Thus, it is still premature to express a judgement on the approaches adopted

---

10 In Europe we can remind the landscape maps and the *landscape character assessments* introduced in the UK following the ELC.

11 The contents and the technical standards of implementation of the development plans adopted by cities and towns must be conformed to the contents and to the technical standard of implementation of the Regional Landscape Plan.

12 Italian: *Legge Urbanistica Nazionale* (LUN), L.1150/1942

13 See the European Landscape Convention: eighth sentence of the introduction, art. 5, art. 6 letters C and D, point 23 of the explanatory report. In the national sphere: L.D. 42/2004 art. 7, 133 and 144, as modified by: L.D. 156/2006, L.D. 157/2006, L.D. 62/2008, L.D. 63/2008, L.D. 86/2009, D.M. 87/2009, D.P.R. n.139/2010.

In particular, the article 133 of the L.D. 42/2004 establishes that the policies on cultural heritage and landscape and the guidelines of landscape planning have to be draft considering the studies and proposals provided by the national and the regional Observatories for the quality of landscape. They are offices established with MiBACT's decree, as well as by the Observatories established with the same goals in each region (subparagraph 1). The regional structures in charge of the participation, often called "Landscape Observatories", are characterized on the one hand by the variety of the activities conducted and on the other hand by the uniformity of the approaches implemented (De Marchi, 2009).

by the Regions in landscape planning. Nonetheless, some experiences like the community maps realised in Apulia for the RLP and the eco-museums in Tuscany, which have become local Observatories of the landscape, have proved the existence of a long tradition in urban planning and the involvement of the academic environment as two important factors for a successful regional planning (Usai, 2016).

7.3 Landscape planning in Sardinia[14]

The landscape planning was introduced in Sardinia by the *Galasso Law* in the Mid-1980s; it produced fourteen Landscape Territorial Plans (LTPs), in addition to the "*Molentargius Monte Urpinu*" LTP already existing[15]. After the *Galasso Law* the regional law on town and country planning returned on the adoption of a regional plan for landscape[16].

Between 2001 and 2003 (Falqui, 2011, pp. 11-13), the Sardinian regional administrative court quashed all the LTPs approved with law n. 431/1985[17]. Consequently, in 2004 the idea of a regional plan for landscape was re-taken by the Sardinia President Renato Soru with the '*salva coste*'[18] law and in 2006 the Sardinia RLP was approved.

The Sardinia plan was the first RLP to be adopted after the L.D. 42/2004 entered into force. The RLP identifies twenty-seven landscape units located along the coast, whose safeguard has priority over the interior areas of the island; their identity features and peculiarities are examined according to three levels of reading: environmental, historical-cultural and residential.

With the L.8/2004, adopting the guidelines of the L.D.42/2004, Sardinia Region established an Observatory for the Planning and Quality of the Landscape[19] allowing the participation of local communities in the safeguard and valorisation of heritage sites and landscapes (Campus and Ercolini, 2010, p. 15).

---

14 The following paragraph is a re-elaborated version of the contributions drawn and discussed in three different international conferences: *Regional planning and territorial competitiveness: the role of Identity Heritage. The case of the Sardinian Region* (Colavitti et al., 2013a); *Debating the Landscape. The Role of Landscape Regional Planning Between Local Community Identity Project and Institutional Landscape Idea* (Colavitti *et al.*, 2013b); *Distinctive and Innovative Regions: Urban Resilient Communities and Regional Policies. The Sardinia Case Study* (Colavitti *et al.*, 2014).

15 Approved according to the L.1497/1939.

16 Regional Law n. 45/1989.

17 Except for the LTP n.7 of Sinis region, which is still in force.

18 Regional law n.8/2004 titled "Urgent measures for the safeguard of the regional territory and for landscape planning".

19 The task of the regional observatory are to periodically update and implement the RLP, to form and develop a culture of landscape, to monitor the quality of changes.

Despite the Sardinia RLP was built to promote an active dialectic with the different stakeholders, actually, the population, probably not only in Sardinian region, was found unprepared to express aware and relevant opinions regarding the choices of the plan (Ercolini *et al.*, 2010, p. 103).

In 2009 the election of a new regional government marked a turning point in landscape planning. New programs on the energy efficiency of buildings are supported (*Piano Casa Sardegna*). A control of partial consistency with the Sardinia RLP is introduced[20]. Finally, the revision and update of the Sardinia RLP started[21]. At this purpose, the regional government launched a participatory process named *Sardegna Nuove Idee* (Sardinia New Ideas). At operative level, the participation took place through workshops involving the stakeholders but also through: *on line* communication; a GIS *open source*[22]; the funding of pilot projects, called "integrated landscape programs", similar to those present in the RLPs of Tuscany and Apulia regions[23] (Table 32).

At the end of the participatory process the Region drew the guidelines for the revision of the Sardinia RLP and its application to the internal areas[24] - not considered by the 2006 Sardinia RLP (RAS, 2012, p. 15).

Table 32. Participation tools in the landscape planning of Sardinia Region

Planning process	Participation tools of Sardinia Region
Drafting stage	*Sardegna Geoblog* Regional Observatory on landscape planning *Sardegna Nuove Idee* process Prize for the Integrated landscape programs
Implementation stage	*Sardegna Geoblog* planning and landscape quality Regional Observatory on landscape planning Prize for the Integrated landscape programs

Source: elaboration by the author

20 This procedure allows the Municipalities whose urban plan for the historical centre is under revision to overcome the stringent rules of safeguard of the RLP allowing the implementation of strategic projects in the historical centres.

21 L.R. 4/2009 Paragraph II.

22 http://webgis.regione.sardegna.it/sardegnageoblog/?page_id=4

23 See RAS (2012), pp. 15, 22-23.

24 In that document it is highlighted the importance of *Sardegna Nuove Idee* to build up a shared idea of Sardinia Landscape. Nevertheless the poor participation of the local communities to the whole process, which is probably related to a lack of information on the objectives and on the real acceptance of the petitions by the political decision-makers, reduced *Sardegna Nuove Idee* to a unilateral activity of dissemination on the regional landscape planning.

On the basis of these guidelines the regional government worked on a "new" RLP, for which the strategic environmental impact assessment was initiated.

In May 2013 the Sardinia Region and the MiBACT signed a memorandum of understanding on the mapping of the regional heritage assets and on the technical standards of implementation for cultural heritage and landscape within the "new" RLP.

The "new" Sardinia RLP was adopted with the Decision of the Regional Government n. 6/18 of 14th February 2014 but the decision caused a constitutional controversy concerning the possibility that the historical settlements were identified as assets of the regional identity heritage and consequently subject to rigorous safeguard rules which limited any form of planning[25]. Similar issues came to light, concerning the strategic environment assessment. Consequently, the regional government revoked the technical standards of implementation of the "new" Sardinia RLP leaving in force the knowledge framework.

Simultaneously, the relationship with MiBACT started again, through the approval of the *Repertorio dei beni paesaggistici e dei contesti identitari* (called also *Repertorio del Mosaico*), updated on 3 October 2014[26]. This document is an inventory of all the heritage assets and landscapes present in the region. They are divided into sections according to a classification established by the 2006 RLP.

The inscription of an historical, cultural or natural site in the inventory takes place after a positive verification of authenticity which is carried out jointly by the Sardinia Region, the Ministry of Culture (MiBACT) and the Municipality which host the site. This process is called *co-pianificazione* (latterly, co-planning).

The inscription of a site to the inventory leads to the adoption of a constraint decree by MiBACT according to the Part II of L.D.42/2004. The decree rules every intervention on the asset and it lists the necessary measures for the safeguard, conservation, maintenance and re-use.

The inventory and the results of the co-planning processes activated at regional level are published on the *Landscape information system of the inventory*[27].

---

25 The new Sardinia RLP was taken up by the State for conflict of powers (appeal 1/2014) in front of the Constitutional Court in relation to the lack of compliance with the co-planning procedure with MiBACT included in the L.D.42/2004 and according the earlier agreements with Sardinia Region.

26 D.G.R. n.10/20 of 28 March 2014.

27 http://www.sardegnageoportale.it/webgis2/sardegnamappe/?map=repertorio2014 *(Landscape information system of the inventory)*

7.4 Integrated landscape programs established by the 2006 Regional Landscape Plan of Sardinia Region

In December 2009 the regional government issued a call for Integrated landscape programs, i.e. five operative pilot projects similar to those present in the RLPs of Tuscany and Apulia regions[28] and co-funded by Sardinia Region[29].

In the call the Integrated landscape programs were defined "as a series of activities and private and/or government interventions, included in a coherent integrated landscape project", i.e. a masterplan. The targets and the objectives of each programme had to include, at least, the integration between conservation and valorisation in cultural heritage and landscape sites.

The call fixed the rule for the competition which involved networks of neighbouring municipalities belonging to the same landscape unit (as defined in RLP) and represented by a leader municipality.

The proposals of Integrated landscape programs were developed by the networks of municipalities starting from the guidelines provided by the Sardinia RLP for each landscape unit. Then the networks detailed the goals, targets and projects of the masterplan (i.e. the *integrated landscape project*)[30].

The award attributed by Sardinia Region to the winning networks of municipalities funded the great part of the interventions contained into the Integrate landscape programs (feasibility studies, preliminary projects or interventions).

The regional government gave importance to the involvement of the local communities adopting a specific criterion in the assessment sheet. The region asked also the competitors to participate at collective meetings during which the proposals were presented and discussed together[31].

---

28 See RAS (2012), pp. 15, 22-23.

29 http://www.regione.sardegna.it/j/v/28?s=1&v=9&c=1425&c1=1425&id=18093

30 The municipalities must co-fund at least the 5 % of the program. The municipalities awarded have to sign a letter of intents which confirms their commitment with the Region for that amount and which obliges them to put the spending in the municipal budget. In this regard, see: *Direzione generale della pianificazione urbanistica territoriale e della vigilanza edilizia, Determinazione n.2886/Pian del 30.12.2009* (Direction General of the land and urban planning and building monitoring, Decree n. 2886/Plan of 30.12.2009).

31 Each program is assessed through a judgment expressed in hundredths with a maximum of 100 points. One part of the assessment comes from the evaluation of the board elected by the Region (maximum 80 points) while the other part comes from the evaluation expressed by the Mayors of the municipalities taking part in the call (maximum 20 points). See Table A and B annex to the Decree n. 2886/Plan of 30.12.2009.

Moreover, the Region analysed carefully the methodologies proposed by the networks of municipalities to implement the programme, if awarded. In particular, the aspects evaluated were:

- *Landscape inclusion*: ability to identify, interpret and specify, according to the local sensibility, the guidelines for the landscape unit provided by the 2006 RLP; ability of the program to govern, through heritage valorisation and landscape management, the future changes in the territory, ensuring the feasibility and compatibility of the potential interventions;
- *relations*: ability of the program to examine the functional, cultural, perceptual and temporal relations among the natural environment, the settlements and the cultural-historical assets;
- *coherence with the landscape unit of the 2006 RLP*: ability of the program to contextualize the masterplan through the study of the weakness and strengths of the territory, the identification of the local identity elements ("the spirit of the city") and the identification of the local landscape features;
- *integration, consistency and feasibility of the interventions of the masterplan*: functional correlation of the interventions and coherence with the project idea underlining the masterplan, i.e. the integrated landscape project;
- *participation process performed for the actuation of the project*;
- *feasibility of the proposal*: presence of the institutional, administrative, technical and procedural conditions required for the implementation and the management of the activities and/or interventions included in the program;
- *composition of the partnership*: coherence and functionality of the partnership in comparison to the objectives of the program.

The form prepared by the regional government for the presentation of the proposals was structured on the following points[32]:

1. *Description of the context*: description of the program context highlighting the strengths and weaknesses of the landscape units involved;
2. *project guidelines of the 2006 RLP*: summary of the guidelines provided by the 2006 RLP for the landscape units involved in the program;

32 General Direction of the land and urban planning and building monitoring, Decree n. 2886/Plan of 30.12.2009, Annex C e Annex D.

3. *objectives and strategies of the integrated program*[33];
4. *significance and coherence of the interventions respect to the program masterplan* (i.e. the integrated landscape project);
5. *summary of the interventions and activities included in the landscape integrated program*: description of the interventions included in the program, the choices and the relations between them, the connections with the masterplan and with the 2006 RLP's guidelines for the landscape units involved;
6. *feasibility of the proposal:* presence of the institutional, administrative, technical and procedural conditions required for the actuation and management of the activities and/or interventions included in the program;
7. *composition of partnership*: description which justifies, in relation to the objectives of the program, the choice of the partners and the role played by each municipality of the network;
8. *participation project carried out drafting the program*: description of the participation project, summary of the results and the objectives outlined, description of the conditions of use of the data collected in the particiaption process in the the design of the masterplan and its inverventions.

On 21 December 2011 the Region offices approved the list with the winning programs and the related budgets[34]. The proposal admitted to the competition were 21 while the projects funded were 12 (Table 33).

The proposals were mainly based on the relationships between the coast and hinterland but some other deal with the unemployment and sustainable tourism (cultural and naturalist itineraries, systems for the slow enjoyment of the landscape).

33 The paragraph answers the following questions: how is the integrated program able to identify, interpret and specify, according to the local sensibilities, the project indications defined by PPR where it rests? How is the integrated program related to the naturalness features and landscape significance, to the territory relations and dynamics of the context where it is included? How does the program promote through development and landscape management interventions, the future changes ensuring the coherence and sustainability of land transformations?

34 General Direction of the land and urban planning and building monitoring, Decree n.5499/Plan of 21.12.2011 Prot. n.75981

Table 33. The Integrated landscape programs awarded and financed by Sardinia Region

Denomination of integrated product	Leading body of the network	Municipalities belonging to the network
Parco del Tirso: oltre i paesaggi d'acqua (Tirso Park: beyond the water landscapes)	Unione Comuni Bassa Valle del Tirso	Ollastra, Siamaggiore,Siamanna, Siapiccia, Simaxis, Solarussa, Villanova Truschedu, Zerfaliu
Le vie dell'acqua e della terra (earth and water pathways)	Cuglieri	Cuglieri, Bonacardo, Milis, Narbolia, Santu Lussurgiu, Seneghe
Panta Rei	Unione dei Comuni del Sarrabus	Castiadas,Muravera, San Vito, Villaputzu, Villasimus
La costa dei Fenici: servizi e itinerari per lo sviluppo turistico sostenibile (Phoenician's coast: services and itineraries for a sustainable tourist development).	Sarroch	Sarroch, Villa San Pietro, Pula, Domus de Maria, Teulada
Sistema Paesaggio (Landscape system)	San Vero Milis	San Vero Milis, Baratili, San Pietro, Bauladu, Niurachi, Tramatza, Zeddiani
Pedraia	Villanova Monteleone	Villanova, Monteleone, Alghero, Ittiri, Padria, Putifigari, Romana, Thiesi
Sulcis: paesaggi del lavoro (Sulcis: work landscapes)	Portoscuso	Portoscuso, Carbonia, Gonnesa, San Giovanni Suergiu
Il golfo di Oristano: dove la terra abbraccia il mare (the gulf of Oristano: where the land meets the sea)	Oristano	Oristano Arborea, Cabras, Palmas Arborea, Santa Giusta
Paesaggio agrario e litorale: bonifica, protezione, riqualificazione e valorizzazione (agricultural landscape and coast: decontamination, protection and improvement)	Maracalagonis	Maracalagonis, Quartucciu, Quartu S. Elena, Settimo San Pietro, Sinnai, Villasimius
I paesaggi della storia dell'uomo nel golfo di Oristano (the landscapes of human history in the Gulf of Oristano)	Morgongiori	Morgongiori, Arbus, Gonnosfanadiga, Guspini, Marrubiu, Masullas, San Nicolò d'Arcidano, Sirsi, Terralba, Uras

Denomination of integrated product	Leading body of the network	Municipalities belonging to the network
Punta verso la Marmora, circuiti d'Ogliastra tra natura e cultura (Punta verso la Marmora Ogliastra pathways between nature and culture)	Bari Sardo	Bari Sardo, Arzana, Elini, Ilbono, Lanusei, Loceri, Villagrani, Strisaili
Accogliere con cura tra mare e montagna (careful welcome between the mountains and the sea)	Lodé	Lodé, Budoni, Lula, Posada, Siniscola, Torpé

Source: RAS (2011)

7.5 A shared methodology for programs dedicated to cultural heritage and landscape in Sardinia

The procedure adopted by Sardinia Region for the selection and funding of integrated landscape programs follows the common stages of a call for tender: deliberation to contract; adjudication; conclusion and approval of the agreement. The preliminary stage is basic, as the integrated programs are not based on local creative spatial policies but on the guidelines provided by the 2006 RLP for the landscape units involved by the program. These guidelines do not provide studies on the social and economic dimensions locally so this task is performed by the network of municipalities competing. Furthermore, the pre-selection stage (selection of studies of pre-feasibility) as well as the *advisory board* are not present.

Finally, the participation to the call is addressed by networks of municipalities which need not to be formalized as a "working group"" after the award. This means that, after the funding, the interventions can be immediately implemented. During this stage, the Region gets in touch with the leader municipality which is the responsible of the re-distribution of funds within the network[35] (Table 34). In this perspective, the methodology adopted by Sardinia Region for the selection of the Integrated landscape programs appears very different from the methodology we propose for the construction of the advanced cultural districts[36].

The methodology established by the Region for the draft of the programs' masterplan, i.e. the *integrated landscape project*, by the competitors has, instead, many points in common with the methodology we illustrated in chapter 6. Both of them started with the analysis of the local resources and relationships; both use the SWOT analysis to determine the strategic posi-

35 General Direction of the land and urban planning and building monitoring, Note Prot. n.24155/Plan 09/07/2010 – Explanations about the most common questions.

36 Which, it must be mentioned, it follows the excellent one adopted by Fondazione Cariplo for the cultural districts in Lombardy (IT).

tioning of territory in question (*benchmarking*) and the GOPP methodology to define the *masterplan* (definition of the problems, specific and general objectives, activities and subjects potentially interested); both pay attention to the *governance*, to the harmonization of the partnership, to the choice of the suitable management form and to the definition of the economic and financial plan (Table 34).

In this case, the methodologies adopted for the draft of the landscape integrated programs show only slight differences which concern: the resources and the relations examined in the first stage (there is only a description of the tangible resources and of the spatial and institutional relationships); the choice of the urban planning tools through which the interventions described in the landscape integrated project can be implemented; the meaningfulness of the monitoring system and the assessment related to the communication plan (Table 34). This last feature is counterbalanced by a great attention to participation which represents the real distinctive feature of the landscape planning.

Table 34. Comparison between the two methodological approaches: application of the framework

Iter for the construction of advanced cultural districts	*Iter* for the draft of the Integrated programs for landscape by Sardinia Region	Drafting of the feasibility study	Drafting of the integrated landscape projects (masterplan)
A. Preliminary stage	Partially present, reference to the PPR	1. Analysis of the resources	Partially present, no immaterial resources
		2. Analysis of the relationships	Partially present, only territory and institutional relations
B. Call for tenders for the selection of pre-feasibility studies	Partially present, the studies of feasibility are selected	3. SWOT matrix for the strategic positioning (*benchmarking*)	Present
C. Funding of the feasibility study	Absent	4. Construction of the masterplan (= integrated landscape project) and its macro-goals, general goals, actions, subjects involved	Present
D. Approval of feasibility studies	Partially present, no formalization of the working group	5. Costruction of the *governance*. *5.1 analysis of the management forms provided by the current laws;*	Present
E. Start-up (implementation of interventions of the feasibility study)	Partially present, no support tools as the advisory board	*5.2 choice of the planning instruments to carry out the actions set in the integrated programme of landscape;*	Absent
		5.3 final choice of the management form	Present
		5.4 economic-financial plan, communication plan, definition of the monitoring system	Partially present, slight relevance to the communication plan

Source: elaboration by the author

7.6 The proposed methodology: potentials, limits and future perspectives in research

The analysis gives a further confirmation about the presence of a disciplinary and administrative separation between landscape planning and policies for the cultural districts. This emerges mostly at the national and regional level while, at an operative level, it seems to disappear thanks to the regeneration urban strategies and to a long-lasting research of an integrated approach in town and country planning. However, the case of Sardinia shows that, at sub-regional level, sometimes the differences can persist if the local institutional *milieu* is not able to support its integrate landscape projects by developing a real autonomous and well-defined "cultural planning", especially when the political background changes. This appears clearly both in the analysis of the regional policies for cultural district in Italy and in the analysis of the communication strategies for the cultural districts. It is worth mentioning that in the Italian regions where the creative districts are fully-developed, they are always preceded by a deep reform of the regional cultural policies.

The landscape planning gives an important contribution to the study and management of the spatial, social and economic dynamics of territories and to the participation processes for community involvement. It is clear from the experience of Sardinia Region that the landscape integrated projects, as *masterplans,* are often reduced to a collection of architectural and environmental interventions aimed at: the recovery, maintenance and networking of the cultural and landscape assets; the improvement of the services for the public and the coordination of the actors involved (central government, local bodies, concessionaires, managers, tourism operators, etc.). Briefly, the Sardinia landscape integrated projects replicated the "project-based" logic of the traditional cultural districts instead of the "program-based" logic of the advanced cultural districts. This reduces the chance to take part in the urban regeneration strategies and in the complex programs on cultural heritage and landscape based on the creative spatial policies.

In this respect, the methodology we propose, seems to offer some useful elements for their development: a theoretical model of reference for the creative district and for the advanced cultural district, a procedure for their construction, a "tool-box" of techniques and indicators to be used in this process and, finally, their potential applications in the city.

The ways in which this knowledge can be applied to the real case are still to be investigated: from the production of quantitative data to be shared and compared to the creation of a proper cartography for the creative spatial policies applied to cultural heritage and landscape. In the future, it is advisable to investigate in these directions.

Bibliography

Adorno, T.W. and Horkheimer, M. (1947). *Dialektik der Aufklärung: Philosophische Fragmente*. Amsterdam: Querido

Ahearne, J. (2009). Cultural policy explicit and implicit: a distinction and some uses, in *International Journal of Cultural Policy*, 15(2), pp. 141-153

Alberti, F.G. and Giusti, J.D. (2009). Alla ricerca dei distretti cultruali: Un analisi critica della letteratura, in *Liuc paper*, Serie management ed economia della cultura 2, n.229/2009, pp. 1-31

Alfken, C. and Broekel, T. and Sternberg, R. (2015). Factors Explaining the Spatial Agglomeration of the Creative Class: Empirical Evidence for German Artists, in *European Planning Studies*, 23(12), pp. 2438-2463

Amabile, T. M. (1983). *The social psychology of creativity*. New York: Springer.

Amabile, T.M. (1998). How to kill creativity, in *Harvard Business Review*, September-October, pp. 77-87

Amari, M. (2006). *La progettazione culturale. Metodologia e strumenti di cultural planning*. Milano: Franco Angeli

Anastasia, B. and Rullani, E. (1981). *La nuova periferia industriale: Saggio sul modello veneto*. Venezia: Arsenale

ANCE and Ambrosetti (2005). *La città dei creativi: visioni e progetti*, rapporto di ricerca, 9 vol., Milano: The European House – Ambrosetti

ANCI (2010). *NOTA DI LETTURA al Federalismo Demaniale: Decreto legislativo 28 maggio 2010, n.85* [on line] Available at: http://www.federalismodemaniale.anci.it/allegati/Nota_lettura_D_Lgs_85.pdf [Last access: 14 May 2014]

ANCI (2012). *Piano nazionale per le Città*, Vademecum per la presentazione delle Proposte [on line] Available at: http://www.anci.it/index.cfm?layout=dettaglio&IdDett=37935 [Last access: 12 May 2014]

Anderson, B. (1983). *Imagined Communities: Reflections on the Origins and Spread of Nationalism*. London: Verso.

Andersson, D.E. and Andersson, A.E. and Mellender, C., eds (2011). *Handbook of Creative Cities*. London: Edward Elgar

Andres, L. and Chapain, C. (2012). The Integration of Cultural and Creative Industries into Local and Regional Development Strategies in Birmingham and Marseille: Towards an Inclusive and Collaborative Governance?, in *Regional Studies*, 47(2), London: Routledge, pp. 161-182

Anholt, S. (2011). Competitive Identity, in Morgan, N. and Pritchard, A. and Pride, R., eds., *Destination brands. Managing place reputation*, Oxford: Elsevier, pp. 21-31

Antoldi, F. (2006), Tra tradizioni locali e competizione globale: introduzione al fenomeno dei distretti industriali, in: Antoldi, F., ed., *Piccole imprese e distretti industriali: politiche di sviluppo in Italia e in America Latina*, Bologna: Il Mulino

Antoniol, M. (2010). *Il federalismo demaniale: Il principio patrimoniale del federalismo fiscale*. Padova: Exeo Edizioni

Archibugi F. (1998). Planning Theory: Reconstruction or Requiem of Planning?. Planning Theory Conference, 2-4 April 1998, Oxford: Brookes University, Oxford Brookes University School of Planning

Argano, L. (2013). Tornare a progettare. La capitale europea della cultura 2019, in Grossi, R., ed., *Una strategia per la cultura una strategia per il paese: IX Rapporto annuale Federculture 2013*, Milano: Il Sole24Ore, pp. 89-114

Asheim, B. and Isaksen, A. (1997). Location, agglomeration and innovation: Towards regional innovation systems in Norway, in *European Planning Studies*, n.5/1997, pp. 299-330

Ashworth, G.J. and Kavaratzis, M. (2009). Beyond the Logo: Brand Management for Cities, in *Journal of Brand Management*, 16(8), pp. 520-531

Astengo, G. (1967). Tutela e valorizzazione dei beni culturali ambientali, in Commissione Franceschini, eds., *Per la salvezza dei beni culturali in Italia. Atti e documenti della commissione di indagine per la tutela e la valorizzazione del patrimonio storico, archeologico, artistico e del paesaggio*, vol.1, Roma: Colombo, pp. 437-504

Bagnasco, A. (1977). *Tre Italie: la problematica dello sviluppo italiano*. Bologna: Il Mulino

Barbati, C. (2008). Le modifiche al Codice dei beni culturali e del paesaggio dopo i decreti legislativi 62 e 63 del 2008. Beni culturali: La valorizzazione: gli artt. 101, 104, 107, 112, 115, 119, in *Aedon –Rivista di arti e diritto on line*, n.3/2008, Bologna: Il Mulino [on line] Available at: http://www.aedon.mulino.it/archivio/2008/3/barbati.htm [Last access: 1 November 2014]

Barbati, C. (2013). Il rilancio dello spettacolo nelle scelte urgenti del decreto "Valore cultura", in *Aedon- Rivista di arti e diritto on line*, n.3/2013 [on line] Available at: http://www.aedon.mulino.it/archivio/2013/3/barbati.htm [Last access: 13 May 2014]

Barth, F. (1981). *Process and Form in Social Life: Selected Essays of Fredrik Barth*, Volume I. London: Routledge

Bathelt, H. and Malmberg, A. and Maskell, P. (2004). Clusters and knowledge: Local buzz, global pipelines and the process of knowledge creation, in *Progress in Human Geography*, 28(1), pp. 31-56.

Bauman, Z. (1998). On Glocalization: Or Globalization for Some, Localization for Some Others, in *Thesis Eleven*, n.54/1998, pp. 37-49

Bauman, Z. (2005). *Liquid Life*. Cambridge: Polity Press

Becattini, G., ed. (1975). *Lo sviluppo economico della Toscana con particolare riguardo all'industrializzazione leggera*. Firenze: IRPET- Guaraldi

Becattini, G. (1979). Dal "settore" industriale al "distretto industriale": Alcune considerazioni sull'unità di indagine dell'economia industriale, in *Rivista di Economie e Politica Industriale*, n.1/1979

Becattini, G. (1989). *Mercato e forze locali: Il distretto industriale*. Bologna: Il Mulino

Becattini, G. (2002). Dal distretto industriale marshalliano alla "distrettualizzazione italiana". Una breve critica, in Quadrio Curzio, A. and Fortis, M., eds., *Complessità dei distretti industriali: Dinamiche, modelli, casi reali*, Bologna: Il Mulino

Becattini, G. (2004). The industrial district as a creative milieu, in Becattini, G., ed., *Industrial district: A new approach to Industrial Change*, Cheltenham: Edward Elgar, pp. 34-47

Bellandi, M. (1987). La formulazione originaria, in Becattini, G., ed., *Mercato e forze locali*, Bologna: Il Mulino, pp. 49-67

Bellandi, M. (1995). Alcune riflessioni in tema di studio comparato e distretti industriali, in *Sviluppo locale*, n.2-3/1995, pp. 74-91

Bellandi, M. and Sforzi, F. (2001). La molteplicità dei sentieri di sviluppo locale, in Becattini, G. and Bellandi, M. and Dei Ottati, G. and Sforzi, F., eds., *Il caleidoscopio dello sviluppo locale*, Torino: Rosenberg & Selllier, pp. 41-63

Bellini, N. and Loffredo, A. and Pasquinelli, C. (2010). Managing Otherness: the political economy of place images in the case of Tuscany, in Ashworth, G. and Kavaratzis, M., *Towards Effective Place Brand Management. Branding European Cities and Regions*, Cheltenham (UK): Edward Elgar, pp. 89-115

Beltram Niessen, M. (2007). Città creative: una rassegna critica sulla letteratura e sulle definizioni, in *Working paper UrbEur PhD*, n.10/2007, Milano: University of Milano-Bicocca [on line] Available at: http://boa.unimib.it/ [Last access: 30 March 2012].

Belussi, F. (2011). Deconstructing Creativity: Entrepreneurs, Individual Talents, and Social Networks, in Belussi, F. and Staber, U., eds., *Managing Networks of Creativity*, London-New York: Routledge, pp. 3-29

Belussi, F. and Staber, U., eds (2011). *Managing Networks of Creativity*, London-New York: Routledge, pp. XV-XVIII

Benevolo, L. (1963). *Le origini dell'urbanistica moderna*. Roma: Laterza

Bennet, T. (2001). *Differing Diversities: Cultural Policy and Cultural Diversity*. Strasbourg: Council of Europe

Bianchini, F. (1990). Urban renaissance? The arts and the urban regeneration process, in Macgregor, S. and Pimlott, B., eds., *Tackling the inner cities: The 1980s Reviewed, Prospects for the 1990s*, Oxford: Oxford University Press

Bianchini, F., (2001). Cultural planning for sustainable development: an overview of emeging professions, paper presented at the Conference "Cultural Policy and Planning Research Unit", Nottingham Trent University.

Bianchini, F. and Fisher, M. and Montgomery, J. and Worpole, K, (1988). *City centres, city cultures: the role of the arts in the revitalization of towns and cities*. Manchester: The Centre for Local Economic Development Strategies

Bianchini, F. and Parkinson, M. (1993). *Cultural Policy and Urban Regeneration: The West European Experience*. New York: Manchester University Press

Boari, C., ed. (2010). *Dinamiche evolutive nei cluster geografici di imprese*. Bologna: Il Mulino

Bocci, C. and Ferrari, R. (2013). Gestione innovativa del settore culturale: forme di parternariato pubblico-privato e possibili strumenti di governance strategica, in Grossi, R., ed., *Una strategia per la cultura, una strategia per il paese: IX Rapporto annaule Federculture 2013*, Milano: il Sole24Ore, pp. 215-228

Bodo, C. and Spada, C., eds. (2013). *Rapporto sull'economia della cultura in Italia 1990-2000*. Bologna: Il Mulino

Bonacci, I. (2007). *Lo Sviluppo organizzativo dei distretti industriali: Il ruolo dell'ICT*. Milano: Franco Angeli

Bondonio, P. and Debernardi, L. (2006). Torino, le valli olimpiche e le eredità dei XX Giochi invernali. Un punto di vista, in Bondonio, P. and Dansero, E. and Mela, A., eds., *Olimpiadi oltre il 2006*, Roma: Carocci, pp. 27-64

Borrup, T. (2015). Creativity in Urban Placemaking: Horizontal Networks and Social Equity in Three Cultural Districts, PhD Thesis, Antioch University – PhD Program in Leadership and Change, AURA Dissertations & Theses, Paper n.233 [on line] Available at: http://aura.antioch.edu/etds/233 [Last access: 12 June 2016]

Boschma, R.A. (2005). Proximity and Innovation: A Critical Assessment, in *Regional Studies*, 39 (1), pp. 61-74

Bossi, G. and Bricco, P. and Scellato, G. (2006). *I distretti del futuro*. Milano: Il Sole 24 Ore

Bourdieu, P. (1985). De la règle aux strategies, in *Terrain*, n.4/1985, pp. 93-100

Bourdieu, P. and Wacquant, L. J. (1992). *Réponses pour une anthropologie reflexive*. Paris: Seuil

Braudel, F. (1969). *Écrits sur l'histoire*. Paris: Flammarion

Brusco, M. (1980). Il modello Emilia: disintegrazione produttiva e integrazione sociale, in *Problemi della transizione*, n.5/1980, pp. 86-105

Byrne, T. (2012). The Creative City and Cultural Policy: opportunity or challenge?, in *Journal of the Department of Cultural Policy and Management*, n.6/2012, London: City University London, pp. 52-78

Buscema, M. and Carlei, V. and Nuccio, M. and Sacco, P.L. (2006). Lo scenario dei distretti industriali attraverso l'analisi delle similarità multidimensionali, Atti della XXVII Conferenza Italiana di Scienze Regionali, Pisa, 12-14 October 2006 [on line] Available at: http://www.aisre.it/ [Last access: 4 September 2014]

Bussi, F.(2001). *Progettare in parternariato: Guida alla conduzione di gruppi di lavoro con il metodo GOPP*. Milano: Franco Angeli

Bussi, F. (2002), "Project Cycle Management e GOPP: principi fondamentali e metodologia di progettazione" e "La tecnica di valutazione ex-ante basata sul Quadro Logico", materiali di approfondimento del Seminario *Progettazione e valutazione di progetti con il Quadro Logico* [on line] Available at: http://db.formez.it/fontinor.nsf/98d556210bdcd5d7c12573540033f844/88B06D764DF054EEC125709D003DC785/$file/Bussi_Progettazione+e+valutazione+di+progetti+con+il+Quadro+Logico.pdf [Last access: 30 December 2014]

Camagni, R. (1999). The city as a *milieu*: applying GREMI's approach to urban evolution, in *Revue d'Economie Régionale et Urbaine*, n.3/1999, pp. 591-606

Camagni, R. (2008). Regional Competitiveness: Towards a Concept of Territorial Capital, in Camagni, R. and Capello, R. and Chizzolini, B. and Fratesi, U., eds., *Modelling Regional Scenarios for the Enlarged Europe*, Berlin: Springer, pp. 33-48.

Camagni, R. (2009). Per un concetto di capitale territoriale, in Borri, D. and Ferlaino, F., eds., *Crescita e sviluppo regionale: stumenti, sistemi, azioni*, Milano: Franco Angeli

Camagni, R. (2012). Verso una riforma della governance territoriale. Area vasta e controllo della rendita fondiaria e immobiliare, in *Quaderni del Territorio*, n.2/2012, pp. 39-62 [on line] Available at: http://www.storicamente.org/ [Last access: 13 May 2014]

Camagni, R. and Capello, R. (2002), Milieux Innovateurs and Collective Learning: From Concepts to Measurement, in Acs, Z. and de Groot, H. and Nijkamp P., eds., *The Emergence of the Knowledge Economy: A Regional Perspective*, Berlin: Springer Verlag, pp. 15-45

Capello, R. (2004). *Economia Regionale*. Bologna: Il Mulino

Camerlengo, Q. (2013). Analisi del quadro normativo-istituzionale per i distretti culturali, in Francesconi, A. and Cioccarelli, G., eds., *Organizzare i distretti culturali evoluti*. Milano: Franco Angeli, pp. 169-242

Cammelli, M. and Della Torre, S. and Hinna, A. and Seddio, P., eds., Fondazione Cariplo, ed. (2007). Come realizzare uno studio di fattibilità operativa per la progettazione di un distretto cuturale: Linee guida [on line] Available at: http://www.fondazionecariplo.it/static/upload/fc_/fc_distrettilinee_web.pdf [Last access: 4 September 2014]

Campus, E. and Ercolini, M. (2010). *Progettare il Paesaggio per sistemi di relazioni. Osservatorio della Pianificazione Urbanistica e della Qualità del Paesaggio*. Olbia: Editrice Taphros

Canino, P. and Cima, S. and Della Torre, S. and Gazzero, L. and Rubini, A. (2013). Apprendere lezioni dalla progettazione e dall'attuazione dei distretti, in Barbetta, G.P. and Cammelli, M. and Della Torre, S., eds., *Distretti culturali: dalla teoria alla pratica*, Bologna: Il Mulino, pp. 189-216

Cariani, R. (2010). *Ecodistretti 2009: made green in Italy: le politiche ambientali dei sistemi produttivi locali e dei distretti industriali*. Milano: Franco Angeli

Carta, M. (1996). *Pianificazione territoriale e urbanistica: Dalla conoscenza alla partecipazione*. Palermo: Medina

Carta, M. (1999). Il valore del patrimonio culturale attraverso l'interpretazione delle carte edelle convenzioni internazionali, in Carta, M., *L'armatura culturale del territorio: Il patrimonio culturale come matrice di identità e strumento di sviluppo*, Milano: Franco Angeli, pp. 89-106

Carta, M. (2002). *L'armatura culturale del territorio: il patrimonio culturale come matrice d'identità e strumento di sviluppo*, II ed. Milano: Franco Angeli

Carta, M. (2004). Strutture territoriali e strategie culturali per lo sviluppo locale, in *Economia della Cultura*, n.1/2004, pp. 39-56

Carta, M. (2007). *Creativity city. Dynamics, innovations, actions.* Barcellona: List

Carta, M. (2009). Creative City 3.0. New scenarios and projects, in *Monograph. it*, n.1/2009

Cartei, G. F. (2008). Codice dei beni culturali e del paesaggio e Convenzione europea: un raffronto, in *Aedon – Rivista di arti e diritto on line*, n.3/2008, Bologna: Il Mulino [on line] Available at: http://www.aedon.mulino.it/archivio/2008/3/barbati.htm [Last access: 1 Novembre 2014]

Casavola P. and Trigilia, C., eds. (2010). *La nuova occasione. Città e valorizzazione delle risorse locali.* Roma: Donzelli

Casoni, G. (2011). Il ruolo dei territori nei processi di innovazione, in Casoni, G. and Fanzini, D., *I luoghi dell'innovazione: Complessità management progetto*, Santarcangelo di Romagna: Maggioli Editore, pp. 51-70

Castells, M. (2001). *Galassia Internet.* Milano: Feltrinelli

Castiglioni, B. (2009). Aspetti sociali del paesaggio: schemi di riferimento, in Castiglioni, B. and De Marchi, M., eds., *DI CHI È IL PAESAGGIO? La partecipazione degli attori nella individuazione, valutazione e pianificazione*, Padova: CLEUP, pp. 73-86

Cerrato, D. (2008). I distretti industriali: elementi costitutivi, prospettive d'analisi e aspetti evolutivi, in Partnership di sviluppo Valor.E, ed., *Valor.E - Valorizzazione delle conoscenze nell'ecosistema formativo*, Napoli: Blado Editori Associati, pp. 45-75 [on line] Available at: http://www.intesasanpaoloformazione.it/images%5CPubblicazioni%5Cvalore_volume_finale.pdf [Last access: 15 July 2014]

Chapuis, M. and Lydon, A. and Brandt-Grau, A., eds. (2009). *Preserving our heritage, Improving our environment*, vol. I "20 years of EU research into cultural heritage", Luxemburg: European Commission –Directorate Environment, Unit Environment Technologies and Pollution Prevention

Chatterton, P. (2000). Will the real Creative City please stand up?, in *City*, 4(3), pp. 390-397

Chiavarino, C. and Gazzerro, L. and Rubini L. (2013). I distratti culturali di Fondazione Cariplo, in Francesconi, A. and Cioccarelli, G., eds., *Organizzare i distretti culturali evoluti.* Milano: Franco Angeli, pp. 107-140

Cicerchia, A., ed. (2010). *Economia della cultura e giovani,* rapporto di ricerca, voll.2, Roma: Edizioni ANCI ComuniCare

Cicerchia, A. (2013). La cultura come oggetto statistico, in Cicerchia, A., *La partecipazione culturale dei giovani in Italia: la musica e l'arte contemporanea*, Milano: Franco Angeli, pp. 26-28

Cirillo, G.P. (2003). La cultura nell'ordinamento giuridico, in *Rass. Cons. Stato*, n.2/2003, pp. 613ss

City Form Lab (2014). Urban Network Analysis: City Form Lab's powerful new tools for the analysis of Urban Activity, in *Evidence Based Design Journal – EBDJ*, August 2014 [on line] Available at: http://ebdjournal.com/blog/urban-design/urban-network-analysis [Last access: 30 September 2014]

Clifton, N. (2008). The "creative class" in the UK: an initial analysis, in *Geografiska Annaler: Series B, Human Geography*, 90(1), pp. 63-82

Clifton, N. (2011). Regional Culture in the Market Place: Place Branding and Product Branding as Cultural Exchange, in *European Planning Studies*, 19(11), pp. 1973-1994

Colavitti, A.M. (2009). *Paesaggio urbano: dalla ricostruzione dei contesti pluristratificati ai piani di recupero della contemporaneità*. Monfalcone (UD): Edicom Edizioni, collana Ambiente e Territorio/Urbanistica e culturale del paesaggio (a cura di G. Deplano)

Colavitti, A.M. (2013). Il Piano di Rinascita della Sardegna. L'innovazione territoriale e le ripercussioni nelle politiche di pianificazione e sviluppo, in Amato, V., ed., *Innovazione, impresa e competitività territoriale nel Mezzogiorno*, Roma: Aracne

Colavitti, A.M. and Usai, A. (2013). Creatività e *place-based development*: buone pratiche e strumenti di rigenerazione creativa per la competitività territoriale, in Colavitti, A.M., *Governance del territorio, Beni Culturali, piano urbanistico. Buone pratiche per la valorizzazione e la competitività,* Firenze: Alinea

Colavitti, A.M. and Usai, A. (2013a). La valorizzazione dei beni immobili pubblici del demanio culturale tra rischi e opportunità. La proposta di un modello di gestione. In: Società Italiana degli Urbanisti, Ed., *Atelier 3. Tutela, gestione e valorizzazione dei beni immobili pubblici*, Atti della XVI Conferenza Nazionale SIU "Urbanistica per una diversa crescita. Aporie dello sviluppo, uscita dalla crisi e progetto del territorio contemporaneo", Napoli 9-10 May 2013, in *PLANUM – The Journal of Urbanism,* n.27/2013, ISSN 1723-0993

Colavitti, A.M. and Usai, A. (2013b). La rigenerazione urbana e la trasformazione dello spazio pubblico: quale ruolo per i beni culturali e paesaggistici?. In: Sbetti, F., Rossi, F., Talia, M., Trillo, C., eds. (2013). *Il governo della città nella contemporaneità. La città come motore di sviluppo*, Atti del XXVIII Congresso dell'INU, Salerno 24-26 October 2013, Urbanistica Dossier, n.4/2013, Roma: INU Edizioni

Colavitti, A.M. and Usai, A. (2014). Federalismo demaniale e autonomie locali: gli strumenti per regolare i rapporti interistituzionali nel trasferimento dei beni costieri appartenenti al ministero della Difesa, in *Aedon*, n.2/2014 [on line] Available at: http://www.aedon.mulino.it/archivio/2014 [Last access: 13 December 2014]

Colavitti, A.M. and Usai, A. (2014a). Partnership building strategy in place branding as a tool to improve cultural heritage district's design. The experience of UNESCO's mining heritage district in Sardinia (Italy), in *Journal of Cultural Heritage Management and Sustainable Development*, 5(2), pp. 151-175

Comune di Genova (2013). Genova Industrie Creative: Stato dell'industria creativa a Genova, Rapporto Finale per il Progetto Europeo *Creative Cities* [on line] Available at: http://www.genovacreativa.it/scenari/genova-industrie-creative-final-versionit [Last access: 30 September 2014]

Comunian, R. and Chapain, C. and Clifton, N. (2014). Creative industries & creative policies: A European perspective?, *City, Culture and Society*, 5(2014), pp. 51-53

Comunian, R. and Sacco, P.L. (2006). NewcastleGateshead: riqualificazione urbana e limiti della città creativa, in *DADI Working Papers*, 2(2006), pp. 1-40

Consiglio Europeo degli Urbanisti (2000). *La Nuova Carta di Atene 1998: I Principi del CEU per la pianificazione delle città*. Firenze: Alinea

Consiglio Europeo degli Urbanisti (2004). *La Nuova Carta di Atene 2003: La visione del CEU per le città del 21° secolo*. Firenze: Alinea

Conzimu, L. and Zoppi, C. (2008). Progetti Integrati di Sviluppo in Sardegna: una valutazione partecipata: Limiti e potenzialità di un processo inclusivo, teesto dell'intervento tenuto in occasione del XI Congresso nazionale dell'Associazione Italiana di Valutazione, Napoli, 28- 30 April 2008 [on line] Available at: http://reteambientale.minambiente.it/wp-content/uploads/2012/06/Regione-Sardegna-Progetti-Integrati-di-Sviluppo.pdf [Last access: 10 December 2014]

Cooke, P. (2001). Regional Innovation Systems, Clusters and the Knowledge Economy, in *Industrial and Corporate Change*, 10(4), pp. 945-974

Cooke, P. and Lazzeretti, L., eds. (2008). *Creative cities, cultural clusters and local economic development*. Cheltenham: Edward Elgar

Cooke, P. and Memedovic, O. (2003). Strategies for Regional Innovation Systems: Learning Transfer and Applications, in *Policy Papers*, Vienna: UNIDO [on line] Available at: http://www.unido.org/fileadmin/user_media/Publications/Pub_free/Strategies_for_regional_innovation_systems.pdf [Last access: 21 November 2013]

Corò, G. and Dalla Torre, R. (2007). Economia della cultura e sviluppo locale, in *Argomenti*, 21(2007), pp. 25-47

Corsaro, A. and Triola, R. (2006). *Legislazione agraria: Contratti, prelazione, imprenditore agricolo*. Milano: Giuffrè di un distretto industriale nel pensiero di Becattini G., in *Economia e politica industriale*, n.107/2000

Costa, P. (2007). Creativity, innovation and territorial agglomeration in cultural activities: the roots of the creative city, in Cooke, P. and Lazzeretti, L. (eds.), *Creative cities, cultural clusters and local development*, Cheltenham: Edward Elgar, pp. 183-210

Costa, P. (2015). Cultural districts and the evolving geographies of underground music scenes: dealing with Bairro Alto's mainstreaming, *Dinamiacet-Iul Working Paper*, 15(2015), DOI: 10.15847/dinamiacet-iul.wp.2015.15

Council of Europe (2000). European Landscape Convention. Florence 20.X.2000, in *European Treaty Series - n.176*, October 2000, Strasburg Cedex: Council of Europe

Council of Europe (2002). *Archeological heritage: The Council of Europe initiatives and the Valetta Convention (1992): Tracking down the traces to understand the present.* Strasbourg Cedex: Council of Europe [on line] Available at: http://www.coe.int/t/dg4/cultureheritage/heritage/resources/Publications/Archaeolgoy_brochure_EN.pdf [Last access: 24 July 2013]

Cozzi, G. (2000). La metamorfosi Csikszentmihalyi, M. (1988). Society, culture, and person: A systems view of creativity, in R. J. Sternberg, ed., *The nature of creativity*, New York: Cambridge University Press, pp. 325-339

Cremaschi, M., ed.(2001). *I programmi integrati, opportunità e vincoli*. Roma: Formez e Donzelli Editore [on line] Available at: http://cremaschi.dipsu.it/about/pubblicazioni/ [Last access: 10 December 2014]

Cremaschi, M. (2002). *Progetti di sviluppo del* territorio: *Le azioni locali integrate in Italia e in Europa*. Milano: Il Sole 24 ORE

Crivello, S. (2010). La città competitiva e sostenibile: alcune riflessioni sul rapporto fra i due discorsi, in Sezione Sociologia del Territorio, Ed., Atti II Conferenza Nazionale Annuale della Sezione "Sociologia del Territorio" dell'Associazione Italiana di Sociologi. Città-campagna: la sociologia di fronte alle trasformazioni del territorio, 25-26 February 2010, Alessandria: Università del Piemonte Orientale [on line] Available at: http://www.sociologiadelterritorio.it/ricerca.php [Last access: 30 March 2012].

Csikszentmihalyi, M. (1996). *Creativity: Flow and the psychology of discovery and invention*. New York: Harper Collins Publishers

Cutaia, L. and Morabito, R. (2012). Ruolo della Simbiosi industriale per la green economy: Uno strumento innovativo per la chiusura dei cicli delle risorse, in *Energia, Ambiente e Innovazione (EAI) - Rivista tecnico-scientifica dell'ENEA*, n. Speciale I-2012, pp. 44-49

Davis, J. (1994). Social Creativity, in Hann, C.M., ed., *When History Accelerates: Essay on Rapid Social Change, Complexity and Creativity*, London: Athlone, pp. 95-100

Debernardi, L., ed. (2005). Distretto della cultura, dello sport e del *loisir* della città di Torino e delle sue valli alpine: Uno sguardo alla teoria e alla letteratura, in *Working papers*, n.1/2005, Torino: Centro di Ricerca sulle Olimpiadi e i Grandi Eventi dell'Università di Torino (*Olympics and Mega Events Research Observatory* – OMERO), pp. 1-17 [on line] Available at: http://www.omero.unito.it/?Working_papers [Last access: 15 July 2014]

De Certeau, M. (1990). *L'invention du quotidien. I Arts de faire*. Paris: Gallimard

De Luca, G. (2012). L'efficacia della pianificazione di area vasta nell'esperienza italiana e comunitaria, Atti del Convegno nazionale INU "La Provincia e la pianificazione intermedia dell'area vasta: nuovi scenari per il governo del territorio", Perugia, 16 April 2012

Della Torre, S., ed., Fondazione Cariplo, Ed. (2006). *I distretti culturali: Studio generale di prefattibilità - Volume 1: Rapporto Conclusivo*. Milano: Politecnico di Milano Dipartimento di Scienza e Tecnologia dell'Ambiente costruito (BEST) [on line] Available at: http://www.fondazionecariplo.it/portal/upload/ent3/1/distretti%20culturali%20rapporto%20conclusivo.pdf [Last access: 30 September 2014]

De Marchi, M. (2009). Partecipazione e Paesaggio, in Castiglioni, B. and De Marchi, M., eds., *DI CHI È IL PAESAGGIO? La partecipazione degli attori nella individuazione, valutazione e pianificazione*, Padova: CLEUP, pp. 123-142

Dematteis, G, (2001). Per una geografia della territorialità attiva e dei valori territoriali, in Bonora, P., ed., *SloT quaderno 1*, Bologna: Baskerville, pp. 11-30

Dematteis, G. and Lanza, C. (1999), La regione geografica, in Conti, S. and Dematteis, G. and Lanza, C. and Nano, F., eds., *Geografia dell'economia mondiale*, Torino: Utet

De Medici, S. (2010). *Nuovi usi per la tutela e la valorizzazione del patrimonio costruito. La privatizzazione dei beni immobili pubblici*. Milano: Franco Angeli

De Medici, S. and Pinto, M.R. (2012), Valorizzazione dei beni culturali pubblici e strategie di riuso, in *TECHNE - Journal of Technology for Architecture and Environment*, n.3, pp. 140-147

De Propris, L. and Lazzaretti, L. (2009). Measuring the decline of a Marshallian industrial District: The Birmingham Jewellery Quarter, in *Regional Studies*, 43(9), pp. 1135-1154

Dei Ottati, G. (1995). *Tra mercato e comunità: aspetti concettuali e ricerche empiriche sul distretto industrial*. Milano: Franco Angeli.

Di Biagi, P. (2009). *I classici dell'urbanistica moderna*. Roma: Donzelli

Dossena, C. (2013). Il distretto culturale e il distretto culturale evoluto tra teoria ed empiria, in Francesconi, A. and Cioccarelli, G., eds., *Organizzare i distretti culturali evoluti*. Milano: Franco Angeli, pp. 243-300

Dossena, C. and Cioccarelli, G. (2013). Dal distretto industriale al distretto culturale evoluto, in Francesconi, A. and Cioccarelli, G., eds., *Organizzare i distretti culturali evoluti*, Milano: Franco Angeli, pp. 27-76

Duxbury, N. (2004). Creative Cities: Principles and Practice, Background Paper F/47 Family Network, August 2004, Ottawa: Canadian Policy Research Networks

Edizel, O. (2013). Mega-events as a place marketing strategy in entrepreneurial cities: İzmir's EXPO 2015 candidacy as a roadmap for hosting EXPO 2020, in *Town Planning Review*, 84 (5), pp. 633-657

Elster, J. (1989). T*he Cement of Society: A study of social order*. Cambridge: Cambridge University Press

Esposito, D. (1996). Carte, documenti eleggi., in Carbonara, G., ed., *Trattato di restauro architettonico*, vol. 4, Torino: Utet, pp. 405-621

Ercolini, M. and Morelli, M. and Natali, C. (2010). Sardegna: il Piano Paesaggistico, tra innovazioni e sfide, in *Ri-Vista ricerche per la progettazione del paesaggio*, January-June 2010

European Commission – Europaid (2001). *Manual Project Cycle Management. General Affairs*. Evaluation Uni: Bruxelles

European Commission (2013). Draft Guidance Fiche Integrated Territorial Investment. Brusseles: European Commission, 27/02/2013 [on line] Available at: http://ec.europa.eu/regional_policy/what/future/pdf/preparation/fiche_integrated_territorial_investment.pdf [Last access: 10 January 2015]

European Commission (2014). *Investment for jobs and growth: Promoting development and good governance in EU regions and cities*, sixth report on economic, social and territorial cohesion, Luxembourg: Publications Office of the European Union [on line] Available at: http://ec.europa.eu/regional_policy/sources/docoffic/official/reports/cohesion6/6cr_en.pdf [Last access: 10 January 2015]

Evangelista, L. (2007). *Bilancio di competenze e orientamento*. Casale Monferrato: Edizioni Sonda

Evangelista, L. (2008). Cos'è il bilancio di competenze?, articolo [on line] Available at: http://www.orientamento.it/indice/cose-il-bilancio-di-competenze/ [Last access: 30 September 2014]

Evans, G. (2001). *Cultural Planning, an Urban Renaissance?*, London: Routledge

Evans, G. (2005). Measure for Measure: evaluating the evidence of culture's contribution to regeneration, in *Urban Studies*, 42(5/6), pp. 959-983

Fabietti, V. (2008), La Pianificazione dei beni dello Stato, in Properzi, P., ed., *Rapporto dal Territorio 2007*, Roma: INU Edizioni, pp. 245-252

Falqui, P.(2011). La vicenda paesistica in Sardegna: dalla Legge Galasso all'annullamento dei PTP (1985-2003), in *Gazzetta Ambiente. Rivista sull'ambiente e il territorio*, Anno 17 n.6, pp. 11-28

Faludi, A. (1973). *Planning Theory*. Pergamon

Federalismo.sspa (2013). *Guida completa al Federalismo demaniale: Le novità per le Regioni e gli enti locali* [on line] Available at: http://federalismo.sspa.it/?p=148 [Last access: 14 May 2014]

Ferilli, G. and Sacco, P.L. and Tavano Blessi, G. (2016). Cities as Creative Hubs: From Instrumental to Functional Values fo Culture-led Local Development, in Girard, L.F. and Baycan, T. and Nijkamp, P., eds., *Sustainable City and Creativity: Promoting Creative Urban Initiatives*, London and New York: Routledge, pp. 245-270

Ferroni, A.M. and Patrignani, S. (2013). Esperienze di partenariato nei programmi di valorizzazione territoriale integrata, in Grossi, R., ed., *Una strategia per la cultura, una strategia per il paese: IX Rapporto annaule Federculture 2013*, Milano: il Sole24Ore, pp. 201-214

Finnel, S. and Santagata, W. 2010. *The Culture Factory: Creativity and the Production of Culture*, London-New York: Springer

Florida, R. (1995). Toward the learning region, in *Futures*, 27(5), London: Routledge, pp. 527-536

Florida, R. (2002). *The Rise of the Creative Class: And How It's Transforming Work, Leisure and Everyday Life*. New York: Basic Books

Florida, R. (2005). *Cities and the Creative Class*. London-New York: Routledge

Florida, R. and Tinagli, I. (2004). *Europe in the creative age*. London: Creativitycitygroup-DEMOS [on line] Available at: http://www.demos.co.uk/publications/creativeeurope [Last access: 30 March 2012]

Florida, R. and Tinagli, I. (2005). *Italy in the Creative Age*, Milan: Creativity-GroupEurope [on line] Available at: http://www.creativeclass.com/rfcgdb/articles/Italy%20in%20the%20Creative%20Age.pdf [Last access: 30 March 2012]

Florida, R. and Mellander, C. and Stolarick, K. (2011). *Creativity and Prosperity: Global Creativity Index*. Toronto: Martin Prosperity Institute [on line] Available at: http://martinprosperity.org/media/GCI%20Report%20Sep%202011.pdf [Last access: 14 January 2014]

Folloni, G. and Gorla, G. (2000). Una modellizzazione del distretto industriale e della suo evoluzione, in *Sviluppo Locale*, 7(13), pp. 33-52

Fondacci, L. (2013). *Gli eco-distretti industriali tra economia e politiche pubbliche*. Firenze: Alinea

Fondazione Cariplo (2007). I distretti culturali,volano economico per il territorio, Bando per l'avvio del progetto di Fondazione Cariplo "Distretti Culturali sul territorio lombardo" [on line] Available at: http://www.fondazionecariplo.it/static/upload/ban/bando-definitivo.pdf [Last access: 30 September 2014]

Foray, D. (2000). *L'Economie de la Connaissance*. Paris: La Découverte Fowler, P.J., ed. (2003). UNESCO World Heritage Papers n°6 – World Heritage Cultural Landscapes 1992-2002, Paris: UNESCO World Heritage Centre [on line] Available at: http://whc.unesco.org/en/series/6/v [Last access: 8 June 2012]

Forgione, L. (2008). *Percorsi di qualità urbana: l'esperienza dei programmi complessi. Approcci, criteri ed esiti*. Roma: Aracne

Francesconi, A. (2013). L'analisi strategico-organizzativa e la progettazione di un distretto culturale evoluto, in Francesconi, A. and Cioccarelli, G., eds., *Organizzare i distretti culturali evoluti*. Milano: Franco Angeli, pp. 117-168

Francesconi, A. and Cioccarelli, G., eds. (2013). *Organizzare i distretti culturali evoluti*. Milano: Franco Angeli

Freeman, C. (1987). *Technology Policy and Economic Performance: Lessons form Japan*. London: Pinter

Fried, R. (1968). Administrative Pluralism and Italian Regional Planning, in *Public Administration*, 46(4), pp. 375-392

Friedmann J. (1993). Pianificazione e dominio pubblico. Dalla conoscenza all'azione, Bari: Dedalo (ed.or. *Planning in the Public Domain: From Knowledge to Action, Princeton*, Princeton University Press, 1987).

Friedmann, J. and Hudson, B. (1974). Knowledge and Action: a guide to planning theory", in *Journal of the AIP*, 40 (1)

Frost-Kump, H. A. (1998). *Cultural Districts: The Arts as a Strategy for Revitalizing Our Cities*. Washington, DC: Americans for the Arts

Gaeta, L. and Mazza, L. and Rivolin, U.J. (2013). *Governo del territorio e pianificazione spaziale*. Novara: Citta Studi Edizioni

Gambino, R. (2003). Progetto e conservazione del paesaggio, in *Ri-Vista Ricerche per la progettazione del paesaggio*, anno 1, numero 0, Firenze: Firenze University Press

Gambino R. (2009). Landscape planning: invarianti e criticità, in Mautone, M. and Ronza, M., eds., *Patrimonio culturale e paesaggio. Un approccio di filiera per la progettualità territoriale*, Roma: Gangemi, pp. 177-182

Garofoli, G. (1983). *Industrializzazione diffusa in Lombardia*. Milano: IReR/ Franco Angeli

Giotta, A.M. and Patruno, A. (2012). "Federalismo demaniale culturale": l'esperienza dell'Agenzia del Demanio, in *NOTIZIARIO dell'Ufficio Studi Segretariato generale MiBAC*, Notiziario XXV-XXVI. 92-97, pp. 28-30

Giordani, P. and Zamparelli, L. (2007). Innovazione e Ricerca nella Teoria Neoclassica, in *Economics Working Papers*, Roma: Università degli Studi di Roma "la Sapienza"

Giovannoni, G. (1931). *Vecchie città ed edilizia nuova*. Torino

Girard, L.F. And Baycan, T. And Nijkamp, P., eds. (2016). *Sustainable City and Creativity: Promoting Creative Urban Initiatives*, London and New York: Routledge

Golinelli, C. M. (2008). *La valorizzazione del Patrimonio culturale: verso la definizione di un modello di governance*. Milano: Giuffrè

Goodall, P. (1995). *High Culture, Popular Culture: The Long Debate*. St Leonards: Allen&Unwin

Governa, F. (2004). *Scegliere, decidere, apprendere. Approcci e criteri nella valutazione dei programmi complessi*, in Regione Piemonte, ed., *Valutare i programmi complessi*, Savigliano: Artistica Editrice

Grandi, R. (2010). Le città creative, in *Il Mulino*, 6(10), pp. 1037-1044

Grogan, D. and Mercer, C. (1995). *The Cultural Planning Handbook: An essential Australian guide*. Sydney: Allen & Unwin

Grossi, R. (2011). Il paese è in ginocchio, diagnostica Roberto Grossi: dunque è arrivato il momento di ripartire, in *Il Giornale dell'Arte*, n.314, Novembre 2011, p. 5

Hall, P. (1998). *Cities in Civilization. Culture, Innovation and Urban Order*. London: Weidenfeld &Nicolson

Hall, P.(2000). Creative cities and economic Development, in *Urban Studies,* 37(4), pp. 639-649

Hankinson,G., 2010. Place branding theory: a cross-domain literature review from a marketing perspective, in Ashworth, G. and Kavaratzis, M., eds., *Towards Effective Place Brand Management. Branding European Cities and Regions*, Cheltenham (UK): Edward Elgar, pp. 15-35

Hannerz, U. (1992). *Cultural Complexity: Studies in the Social Organization of Meaning*. New York: Columbia University Press

Hinna, A. and Seddio, P. (2013). Imprese, risorse e sviluppo: ipotesi e dibattito intorno ai distretti culturali, in Barbetta, G.P. and Cammelli, M. and Della Torre, S., eds., *Distretti culturali: dalla teoria alla pratica*, Bologna: Il Mulino, pp. 21-66

Howard, E. (1898). *To-Morrow: A Peaceful Path to Real Reform*. London: Swann Sonnenschein

Howard, E. (1902). *Garden Cities of Tomorrow*. London: Swann Sonnenschein

Howkins, J. (2001). *The Creative Economy: How People Make Money from Ideas*. London: Penguin

Howkins, J. (2010). *Creative Ecologies: Where Thinking Is a Proper Job*. Queensland: University of Queensland Press

Hutton., T.A. (2016). *Cities and the Cultural Economy*. London and New York: Routledge

Iammarino, S. (2004). On the definition of regional system of innovation (RSI): an application to the Italian case, paper presented at the International Conference on "Regionalisation of innovation policy: options and experiences", German Institute of Economic Research (DIW Berlin), Berlin 4-5 June 2004 [on line] Available at: http://www.diw.de/documents/dokumentenarchiv/17/41807/iammarino.pdf [Last access: 21 Novembre 2013].

ICOMOS (1966). Resolutions on the Regeneration of Historic Urban Sites, Levoca, 20-26 June 1966 [on line] Available at: http://www.icomos.org/publications/93towns7.pdf [Last access: 24 July 2013]

INTELI, (2011). *Creative-based Strategies in Small and Medium-sized Cities: Guidelines for Local Authorities*, Lisboa: INTELI – InteligênciaemInovação, Centro de Inovação [on line] Available at: www.inteli.pt [Last access: 30 September 2013].

ISPRA, Ministero dell'Ambiente e delle tutela del territorio e del mare - MATTM (2012). *Indicazioni metodologiche e operative per il monitoraggio VAS* [on line] Available at: http://www.isprambiente.gov.it/it/temi/valutazione-ambientale-strategica-vas/ispra-e-la-vas [Last access: 10 December 2014]

ISTAT (2009). Classificazione delle attività economiche: Ateco 2007 derivata dalla Nace Rev. 2, *Metodi e Norme*, n.40/2009, Roma: ISTAT [on line] Available at: http://www3.istat.it/dati/catalogo/20090615_00/metenorme09_40classificazione_attivita_economiche_2007.pdf [Ultimo ccesso: 9 May 2014]

Jacobs, J. (1961). *The Death and Life of Great American Cities*. New York: Random House

Jacobs, J. (1969). *The Economy of Cities*. New York: Vintage.

JPI Cultural Heritage – Governing Board (2010). The Joint Programming Initiative on Cultural Heritage and Global Change: a new challenge for Europe, Vision document, 17 June 2010 [on line] Available at: http://www.jpi-culturalheritage.eu/wp-content/uploads/Vision-Document_17-June-20101.pdfb [Last access: 24 July 2013]

Jokilehto, J. (1995). Authenticity: a general framework for the concept, Proceedings of the Nara Conference on Authenticity in relation to the World Heritage Convention, Nara, Japan, 1-6 November 1994, Paris: UNESCO World Heritage Centre – Agency for Cultural Affairs

Jokilehto, J. (2008). Aesthetics in the World Heritage context, in *Values and criteria in heritage conservation,* Proceedings of the International Conference of ICOMOS, ICCROM and Fondazione Romualdo Del Bianco, Florence, 2-4 March 2007, Firenze: Edizioni Polistampa, pp. 183-192

Kavaratzis, M. (2004). From city marketing to city branding: towards a theoretical framework for developing city brands, in *Place Branding*, 1(1), pp. 58-73

Kavoura, A. (2013). *Contemporary Approaches of the Scientific Theory of Place marketing*. New York: Nova Publisher

KEA (2006). *The Economy of Culture in Europe*. Brussels: KEA European Affairs [on line] Available at: http://www.keanet.eu/ecoculture/studynew.pdf [Last access: 24 July 2013]

KEA (2012). *Use of structural funds for cultural projects*, Brussels: KEA European Affairs [on line] Available at: http://www.europarl.europa.eu/RegData/etudes/etudes/join/2012/474563/IPOL-CULT_ET(2012)474563_EN.pdf [Last access: 24 July 2013]

Kern, P. and Montalto, V. (2013). Europa e cultura: quali risorse per il periodo 2014-2020?, in Grossi, R., ed., *Una strategia per la cultura una strategia per il paese: IX Rapporto annuale Federculture 2013*, Milano: Il Sole24Ore, pp. 73-88

Kim, J.Y. (2015). Cultural entrepreneurs and urban re generation in Itaewon, Seoul, in *Cities*, (2015), DOI: http://dx.doi.org/10.1016/

Kim, S.-h. (2015). Belated but grand? The future of public housing in Korea, in *City, Culture and Society,* 5 (2014), pp. 97-105

Lai, F. (2006). *La creatività sociale*, Roma: Carocci Editore

Landry, C. (2000). *The Creative City: A toolkit for urban innovators*. London: Earthscan

Landry, C. (2006a). *The Art of City Making*. London: Earthscan

Landry, C. (2006b). Lineages of the Creative City, in *Research Journal for Creative Cities* (RJCC), 1(1), pp. 15-23

Landry, C. and Bianchini, F. (1995). *The creative city*. London: Demos

Larsson, U., ed. (2001). *Cultures of Creativity: The Centennial Exhibition Of The Nobel Prize*, CantonMA: Science History Publications

Lazzeretti, L. (2008). The cultural districtualisation model, in Cooke, P. and Lazzeretti, L., eds., *Creative Cities, Cultural Clusters and Local Development*, Cheltenham: Edward Elgar, pp. 93-120

Lazzeretti, L. (2011). Culture as a source for growth and change: Some evidence form cultural cluster in Andalusia, in Cooke, P. and Asheim, B. T. and Boschma, R. and Martin, R. and Schwartz, D. and Todtling, F., eds., *Handbook of Regional Innovation and Growth*, Cheltenham: Edward Elgar, pp. 350-362

Lazzeretti, L. (2012). *Cluster creativi per i beni culturali: L'esperienza toscana dalle tecnologie per la conservazione e valorizzazione*. Firenze: Firenze University Press

Lazzeretti, L. (2013). Cultural and creative industries: An introduction, in Lazzeretti, L., ed., *Creative industries and innovation in Europe: concepts, measures and comparative case studies*. London: Routledge

Leontief, W. (1951). *The Structure of America Economy 1919-1929*. New York: Oxford University Press

Leontief, W. (1986). *Input-output Economics*. New York: Oxford University Press

Lévi-Strauss, C. (1983). *Le regard éloigné*. Paris: Plon

Lundvall, B., ed. (1992). *National Systems of Innovation: Towards a Theory of Innovation and Interactive Learning*, London: Pinter

Maggino, F. (2005). Ricerca e individuazione dei gruppi, in Maggino, F., *L'analisi dei dati nell'indagine statistica - Volume 2: l'esplorazione dei dati e la validazione dei risultati*, Firenze: Firenze University Press, pp. 1-22

Mahdjoubi, D. (1997). Schumpeterian economics and the trilogy of "innovation-invention-diffusion" [on line] Available at: http://www.ischool.utexas.edu/~darius/17-Schumpeter-innovation.pdf [Last access: 2 September 2011].

Mariotti, A. (2010). Il distretto come modello intersettoriale di sviluppo del territorio, in *AlmaTourism*, 1(2), pp. 38-54

Markusen, A. (2006). Urban development and the politics of a creative class: evidence from a study of artists, in *Environment and Planning A*, 38 (10), pp. 1921-1940

Marontate, J. and Murray, C. (2010). Neoliberalism in Provincial Cultural Policy Narratives: Perspectives from Two Coasts, in *Canadian Journal of Communication,* 35(2010), pp. 325-343

Marshall, A. (1919). *Industry and trade*. London: Mcmillan

Marshall, A. (1920). *Principles of Economics*. London: Macmillan

Marshall, A. and Marshall Paley, M. (1879). *The Economics of Industry*. London: Mcmillan

Martí-Costa, M. and Pradel i Miquel, M. (2011). The knowledge city against urban creativity? Artists' workshops and urban regeneration in Barcellona, in *European Urban and Regional Studies*, 19(1), pp. 92-108

Mautone, M. (2009). Patrimonio culturale e paesaggio. Dalla conoscenza alla gestione territoriale, in Mautone, M. and Ronza, M., eds., *Patrimonio culturale e paesaggio*, Roma: Gangemi Editore, pp. 5-14

Mele, R. and Parente, R. and Petrone, M. (2008). La *governance* pubblica dei distretti tecnologici, in *Sinergie*, n.77/2008, pp. 81-100

Melina, D., ed. (2011). *Dalla Strategia di Lisbona a Europa 2020: Fra governance e government dell'Unione europea.* Ivrea-Roma: Fondazione Adriano Olivetti

Mercer, C. (1996). By accident or design. Can culture be planned?, in Matarasso, F. and Halls, S., eds., *The Art of Regeneration. Nottingham 1996: Conference Papers Nottingham e Bournes Green*, Nottingham –Lodon: City of Nottingham and Comedia.

Mercer, C. (1997). Mapping regional cultures, in Mules, W. and Miller, H., eds., *Mapping Regional Cultures: Discourses in Social Contexts*, Rockhampton: Central Queensland University, pp. 6-15

Mercer, C. (2008). Bergen and Beyond: the role of municipalities, cities, and city-regions in developing the creative economy, in Mercer, C., *Hands-On: Regional Strategies for Creative Industries in Bergen, Tartu and Turku*, Nordic Innovations Centre (NICe) and City of Tartu, Estonia [on line] Available at: http://www.luts.ee/kjoomla/creativeindustriespublication/publication.pdf [Last access: 12 June 2016]

MiBAC - Direzione Generale per i Beni Archeologici, ed. (2005). *Progetto di definizione di un modello per la realizzazione dei Piani di Gestione dei siti UNESCO.* Roma: MiBAC/Ernst & Young, pp. 189-202 [on line] Available at: http://www.unesco.beniculturali.it/Roma [Last access: 24 October 2014].

MiBAC - Commissione sulla Creatività e Produzione di Cultura in Italia (2007). *Libro Bianco sulla Creatività.* Roma: MiBAC – Ufficio Studi [on line] Available at: http://www.ufficiostudi.beniculturali.it/mibac/export/UfficioStudi/sito-UfficioStudi/Contenuti/Pubblicazioni/Volumi/Volumipubblicati/visualizza_asset.html_1410871104.html [Ultimo accesso. 26 March 2014]

MiBAC and Istituto Tagliacarne (2007). *Le attività economiche collegate alla valorizzazione del patrimonio culturale.* Roma: Istituto Tagliacarne [on line] Available at: http://www.tagliacarne.it/P42A172C171S0/Le-attivita-economiche-collegate-alla-valorizzazione-del-patrimonio-culturale.htm [Last access: 30 April 2014].

MiBAC - Commissione sulla Creatività e Produzione di Cultura in Italia (2009). *Libro bianco sulla creatività.* Roma: MiBAC [on line] Available at: http://www.ufficiostudi.beniculturali.it/mibac/export/UfficioStudi/sito-UfficioStudi/Contenuti/Pubblicazioni/Volumi/Volumi-pubblicati/visualizza_asset.html_1410871071.html [Last access: 10 May 2014]

Ministro per la Coesione Territoriale - Ufficio Stampa (2014). Progetto "Cultura in Movimento": i Ministri Trigilia e Bray hanno incontrato i sindaci delle città partecipanti, Comunicato stampa del 30 January 2014 [on line] Available at: http://www.coesioneterritoriale.gov.it/cultura-movimento-incontro/ [Last access: 10 May 2014]

Montella, M., ed. (2015). Distretti culturali: esperienze a confronto, in *Il capitale culturale. Studies on the Value of Cultural Heritage, Supplementi,* 3(2015)

Montgomery, J., (2003). Cultural Quarters as Mechanisms for Urban Regeneration: Part 1: Conceptualising Cultural Quarters, in *Planning, Practice & Research*, 18 (4), pp. 293-306

Mould, O. and Comunian, R. (2015). Hung, Drawn and Cultural Quartered: Rethinking Cultural Quarter Development Policy in the UK, in *European Planning Studies*, 23(12), pp. 2356-2369

Mumford, L. (1960). *The City in History: Its Origins, Its Transformations, and Its Prospects*. New York: Harcourt, Brace and World

Murdoch, J. and Grodach, C. and Foster, N. (2016). The Importance of Neighborhood Context in Arts-Led Development: Community Anchor or Creative Class Magnet?, in *Journal of Planning Education and Research*, 36(1), pp. 32-48

National Park Service, 1995. How to Evaluate the Integrity of a Property, in *National Register Bullettin*, Washington: U.S. Government printing office, pp. 44-48 [on line] Available at: http://www.nps.gov/nr/publications/bulletins/pdfs/nrb15.pdf [Last access: 12 February 2014]

Niglio, O. (2009). Conservazione e attualità degli ambienti antichi. Dalla Carta di Atene del 1931 alla Carta di Betlemme del 2008, in *EdA – Esempi di Architettura*, n.7/2009, Padova: Editore Il Prato, pp. 31-41. [on line] Available at: http://www.esempidiarchitettura.it/ebcms2_uploads/fckeditor/file/Carta%20Bethlehem%202008.pdf [Last access: 8 June 2012]

Noonan, D.S. (2013). How US cultural districts reshape neighbourhoods, in *Cultural Trends,* 22(3-4), pp. 203-212

Nuccio, M and Ponzini, D. (2016). What does a cultural district actually do? Critically reappraising 15 years of cultural district policy in Italy, in *European Urban and Regional Studies*, (2016), DOI 10.1177/0969776416643749

O' Connor, J. and Shaw, K. (2014). What next for the creative city?, in *City, Culture and Society* (in press), DOI: 10.1016/j.ccs.2014.05.010

OECD, 2006. *OECD Territorial Reviews: Competitive Cities in the Global Economy*. Paris: OECD

Palmi, P. (2013). *Le fabbriche della creatività: Un'analisi organizzativa dei distretti evoluti*. Milano: Franco Angeli

Pareja-Eastaway, M. and Pradel i Miquel, M. (2015). Towards the Creative and Knowledge Economies: Analysing Diverse Pathways in Spanish Cities, in *European Planning Studies*, 23(12), pp. 2404-2422

Peck, J. (2012). Recreative City: Amsterdam, Vehicular Ideas and the Adaptive Spaces of Creativity Policy, in *International Journal of Urban and Regional Research*, 36(3), pp. 462-85

Peck, J., Theodore, N. and Brenner, N. (2009). Postneoliberalism and its malcontents, in *Antipode,* 41(s1), pp. 94-116

Perrone, V. (2004). Oltre l'orizzonte dei distretti, in *Economia & Management*, n.5/2004, pp. 3-11

Pinheiro, R. and Hauge, E. (2014). Global scripts and local translations: The case of cultural and creative industries (CCIs) in Norway, in *City, Culture and Society*, 5(2014), pp. 87-95

Ponzini, D. (2008). *Il territorio dei beni cultural: Interpretazioni strategiche del processo di privatizzazione dei beni e delle attività culturali in Italia*. Roma: Carocci Editore

Ponzini, D. (2016). Cultural policy-making by netwroking, in Hristova, S. and Dragićević Šešić, M. and Dragi Evi, m. and Duxbury N., eds., *Culture and Sustainability in European Cities: Imagining Europolis*, London and New York: Routledge, pp. 100-109

Ponzini, D. and Gugu, S. and Oppio, A. (2014). Is the concept of the cultural district appropriate for both analysis and policymaking? Two cases in Northern Italy, in *City, Culture and Society*, n.5/2014, pp. 75-85

Porrello, A. (2006). *L 'arte difficile del cultural planning*. Venezia: Università IUAV - Dipartimento di pianificazione [on line] Available at: http://www.iuav.it [Last access: 30 March 2012].

Porter, M.E. (1990). *The Competitive Advantage of Nations*. New York: The Free Press

Porter, M.E. (2003). The Economic Performance of Regions, in *Regional Studies*, 37(6-7), pp. 549-578

Power, D. (2010). *Priority Sector Report: Creative and Cultural Industries*, Europe Innova, European Cluster Observatory, March 2010 [on line] Available at: http://www.clusterobservatory.eu/common/galleries/downloads/CreativeAndCulturalIndustries.pdf [Last access: 12 May 2014]

Pratt, A.C. (2000). Stiky Places? A commentary on the making of 'new media' spaces, paper presented at "Mdia@Ise / USC Annenberg, Managing the Global" Workshop in Media and Communictions at the London School of economics December 2 2000

Pratt, A. C. (2004). The Cultural Economy: A Call for Spatialized 'Production of Culture' Perspectives, in *International Journal of Cultural Studies*, 7(1), pp. 117-128

Pratt, A.C. (2008). Creative cities?, in *Urban Design Journal*, n.106/2008, p. 35

Pratt, A.C. (2011). The cultural contradictions of the creative city, in *City, Culture and Society*, 2(3), pp. 123-130

Pratt, A.C. and Taylor, C. (2006). *A framework for the cultural sector: a report for UIS/UNESCO*. London: BOP

Preite, M., ed. (1998). *La valorizzazione del patrimonio culturale in Toscana: una valutazione di alcune esperienze*. Firenze: Fondazione Michelucci

Priore, R. (2005). Verso l'applicazione della Convenzione europea del paesaggio in Italia, in *Aedon – Rivista di Arte e diritto on line*, n.3/2005, Bologna: Il Mulino [on line] Available at: http://www.aedon.mulino.it/archivio/2005/3/priore.htm [Last access: 7 July 2012)

Putnam, R. D. (1993). *Making democracy work*. Princeton: Princeton University Press Rivas, M. (2011). From Creative Industries to the Creative Place. Refreshing the Local Development Agenda in Small and Medium-sized Towns, in *REDIGE*, 2(2), pp. 131-147

RAS – Direzione generale della pianificazione urbanistica territoriale e della vigilanza edilizia (2011). Determinazione n.5499/PIAN del 21.12.2011 Prot. n.75981, Allegato A [on line] Available at: http://www.regione.sardegna.it/documenti/1_19_20111221130928.pdf [Last access: 10 December 2014]

RAS – Regione Autonoma della Sardegna (2012). Piano paesaggistico regionale: Linee guida, documento in allegato al DOC. n.27/XIV del Consiglio Regionale della Sardegna, 10/07/2012 [on line] Available at: http://www.sardegnaterritorio.it/documenti/6_288_20120809095500.pdf [Last access: 10 December 2014]

RAS – Regione Autonoma della Sardegna, Centro Regionale di Programmazione (CRP) (2012). JESSICA Joint European Support for Sustainable Investment in City Areas Guida per l'attuazione dei progetti finanziati da JESSICA Sardegna a valere sul POR FESR 2007-2013, Vademecum [on line] Available at: http://www.sardegnaprogrammazione.it/documenti/35_84_20120927120016.pdf [Last access: 10 December 2014]

Rivas, M. (2011). From Creative Industries to the Creative Place. Refreshing the Local Development Agenda in Small and Medium-sized Towns, in *REDIGE*, 2(2), pp. 131-147

Rivas, ed. (2012). City Logo: Innovative place brand management, Baseline report _ 2012, report elaborated within the framework of the URBACT CityLogo project [on line] Available at: http://urbact.eu/fileadmin/Projects/CityLogo/documents_media/baseline_citylogo_2012.pdf [Last access: 14 October 2014]

Robertson, R. (1992). *Globalization: Social Theory and Global Culture*. London: SAGE

Robertson, T.S. (1967). The Process of Innovation and the Diffusion of Innovation, in *Journal of Marketing*, Vol. 31 (January, 1967), pp. 14-19.

Roodhouse, S. (2010). *Cultural Quarters: Principles and Practice*. Bristol (UK) and Chicago (USA): Intellect

Ronza, M., (2009). Il distretto culturale nella pianificazione territoriale, in Mautone, M. and Ronza, M., eds., *Patrimonio culturale e paesaggio*, Roma: Gangemi Editore, pp. 201-208

RRA LUR - Regional Development Agency of the Ljubljana Urban Region (2011). Potentials of creative urban regeneration: Spatial distribution of creative industries on Ljubljana Urban Region, Rapporto Finale per il Progetto Europeo *Creative Cities*, Ljubljana: Institute for spatial policies [on line] Available at: http://www.creativecitiesproject.eu/en/doc_download/Other-Documents-file/22-11-2011/Potentials_of_Creative_urban_regeneration.pdf [Last access: 14 October 2014]

Rullani, E. (2002). Il sistema industriale come sistema adattativo complesso, in Quadrio Curzio, A. and Fortis, M., eds.,*Complessità dei distretti industriali: Dinamiche, modelli, casi ideali*, Bologna: Il Mulino

Rullani, E. (2004). *La fabbrica dell'immateriale*. Roma: Carocci

Rushton, M. (2015). Cultural districts and economic development in American cities, in *Poetics*, 49(2015), pp. 20-29

Sacco, P.L., ed., Consorzio Universitario Piceno, Ed. (2007). *Studio di fattibilità per la progettazione di un distretto culturale evoluto nel territorio della provincia di Ascoli Piceno*. Bologna: goodwill

Sacco, P.L. and Blessi, G.T. (2006). Verso un nuovo modello di sviluppo sostenibile: distretti culturali e aree urbane, in *Working paper*, n.6/2006 Venezia: DADI, Università IUAV

Sacco, P.L. and Ferilli, G. (2006). Il distretto culturale evoluto nell'economia post industriale, in DADI *Working papes,* 4(2006), pp. 1-28

Sacco, P.L. and Ferilli, G. and Lavanga, M. (2006). *The cultural district organizational model: a theoretical and policy design approach*. Venice: DADI, Università IUAV.

Sacco, P.L. and Ferilli, G. and Tavano, G. and Nuccio, M. (2013b). Culture as an Engine of Local Development Processes: System-Wide Cultural Districts I: Theory, in *Growth and Change*, 44(4), pp. 555-570

Sacco, P.L. and Pedrini, S. (2003). Il distretto culturale: mito o opportunità?, in *Il Risparmio*, 51(3), pp. 101-155

Saccomani S., (2004), *Programmi complessi: una rilettura delle esperienze*, in Regione Piemonte, ED. *Valutare i programmi complessi*, Savigliano: Artistica Editrice

Sager, T. (2011). Neo-liberal urban planning policies: A literature survey 1990-2010, in *Progress in Planning*, 76(4), London: Elsevier, pp. 147-199

Salvemini, S. (2008). I neodistretti industriali tra nuova cultura e antico territorio, in *Economia & Management*, n.3/2008, pp. 3-10

Santagata, W. (2002). Cultural Districts, Property Rights and Sustainable Growth, in *International Journal of Urbana and Regional Research*, 26(1), pp. 9-23

Santagata, W. (2006). Cultural Districts and Their Role in Developed and Developing Countries, in Ginsburgh, V. and Throsby, D., eds., *Handbook of the Economics of Art and Culture,* Amsterdam: North Holland, pp. 1101-1119

Santagata, W. (2007). *La fabbrica della cultura*. Bologna: Il Mulino

Santagata, W., ed. (2009). *Libro bianco sulla creatività. Per un modello italiano di sviluppo*. Milano: Università Bocconi-Egea

Scandale, L. (2005). Verso *ideapolis*. Cultura e creatività come fattori di competitività della Knowledge-based city, in *Archivio di Studi Urbani Regionali*, 36(84), pp. 107-122

Sciullo, G. (2008). Il paesaggio fra la Convenzione e il Codice, in *Aedon - Rivista di arti e diritto on line*, n.3/2008, Bologna: Il Mulino [on line] Available at: http://www.aedon.mulino.it/archivio/2008/3/barbati.htm [Last access: 24 June 2014]

Scott, A.J. (2006). Creative cities: conceptual issues and policy questions, in *Journal of Urban Affairs*, 28(1), pp. 1-17

Scott, A.J. (2010). Cultural economy and the creative field of the city, in *Geografiska Annaler: Series B, Human Geography*, 9(2), pp. 147-167

Scott, A.J. (2012). *A World in Emergence; Cities and Regions in the Twenty-First Century*. Cheltenham, UK, and Northampton, MA: Edward Elgar

Scott, A.J. (2014). Beyond the Creative City: Cognitive-Cultural Capitalism and the New Urbanism, in *Regional Studies*, 48(4), pp. 565-578

Schumpeter, A.J., 1939. *Business Cycles*, vol.1, New York: McGraw Hill

Scimeni, G. (1979). Voce "Pianificazione", in Treccani Enciclopedia Italiana - IV Appendice (1979) [on line] Available at: http://www.treccani.it/enciclopedia/pianificazione_(Enciclopedia-Italiana)/ [Last access: 28 October 2014]

Seddio, P. (2004). Dalle reti interistituzionali alla costruzione di un sistema di governance pubblica: un primo passo verso forme di distrettualizzazione delle risorse culturali in Italia, in Sibilio Parri, B., ed., *Creare e valorizzare i distretti museali*, Milano: Franco Angeli, p. 113-130

Seddio, P. (2013). *La gestione integrata di reti e sistemi culturali: Contenuti, esperienze e prospettive*. Milano: Franco Angeli

Sen, A. (1994). *La disegualianza*. Bologna: Il Mulino

Sen, A. (2002). *Globalizzazione e libertà*. Milano: Mondadori

Settis, S. (2010). *Paesaggio Costituzione Cemento: La battaglia per l'ambiente contro il degrado civile*. Torino: Einaudi

Schama, S. (1995). *Landscape and Memory*. New York: A.A. Knopf

Sitte, C. (1889). *Der Städtebau nach seinen künstlerischen Grundsätze.* Wien: Graeser

Sjøholt, P. (1999). Culture as a Strategic Development Device: The Role of 'European Cities of Culture with Particular Reference to Bergen, in *European Urban and Regional Studies*, 6(4), pp. 339-347

Soda, G. (1998). *Reti tra imprese*. Roma: Carocci

Sotti F. (2013). Appendice: la fattibilità economico-finanziaria di un progetto di distretto culturale, in Francesconi, A. and Cioccarelli, G., eds., *Organizzare i distretti culturali evoluti*. Milano: Franco Angeli, pp. 309-327

Staber, U. (2011). The Ecological Foundations of Creativity, in Belussi, F. and Staber, U., eds., *Managing Networks of Creativity*, London-New York: Routledge, pp. 30-45

Stern, M.J. and Seifert, S.C. (2008). *Knight Creative Communities Initiative (KCCI): Evaluation: Final Report*. Miami: John S. and James L. Knight Foundation [on line] Available at: http://www.sp2.upenn.edu/siap/docs/knight_creative_communities/final_report.pdf [Last access: 24 January 2014]

Storper, M. (2003). Le economie locali come beni relazionali, in Garofoli, G., ed., *Impresa e territorio*, Bologna: Il Mulino, pp. 169-207

Storto G. (2013), Il Piano Nazionale per le Città: indicazioni procedurali e considerazioni dopo la selezione dei progetti, in *Quaderni di legislazione tecnica*, n.1/2013, pp. 17-21

Sternberg, R.J. (2006). The Nature of Creativity, in *Creativity Research Journal*, 18(1), pp. 87-98

Sternberg, R. J. and Lubart, T. I. (1991). An investment theory of creativity and its development, in *Human Development*, 34(1), pp. 1-31

Stevenson, D. (2014). *Cities of Culture: A global perspective*. Milton Park, Oxon, and New York, NY: Routledge

Sugden, R. (1989). Spontaneous order, in *Journal of Economic Perspectives*, n.3/1989, pp. 85-97

Tabellini, G. (2005). Culture and Institutions: economic development in the regions of Europe, in *CESIFO Working Papers*, Paper n.1492/2005, Munich (Germany): CESifo Group

Tang, W.-S. (2015). Creative industries, public engagement and urban redevelopment in Hong Kong, in *City*, (2015), DOI: http://dx.doi.org/10.1016/j.cities.2015.09.004

Tartaglia, A. (2006). Partenariato pubblico privato negli interventi di valorizzazione dei beni culturali immobiliari: il project financing come opportunità, in Oppio, A. and Tartaglia, A., eds., *Governo del territorio e strategie di valorizzazione dei beni culturali*, Milano: Libreria CLUP, pp. 179-185

Thomas H. and Healy P. (1991). *Dilemmas of Planning Practice*. Aldershot, Hants: Avebury

Throsby, D. (2001). *Economics and culture*. Cambridge: Cambridge University Press

Tomei, G. (2013). Comprendere il cambiamento nei programmi complessi. Necessità e sfide per la valutazione dell'aiuto allo sviluppo dopo la Dichiarazione di Parigi, in *RIV Rassegna Italiana di Valutazione*, n.56-57/2013, pp. 141-170

Torre, C.M. and Di Bitonto, S. and Selicato, M. (2011). Applicazioni della Valutazione ambientale strategica ai piani/programmi regionali e locali: Note metodologiche sulla realtà pugliese, in *AESTIMUM*, n.59/2011, pp. 55-63

Trigilia, C. (1998). *Sociologia economica: Stato, mercato e società nel capitalismo moderno*. Bologna: Il Mulino

Trigilia, C. (2014). Relazione sull'attività svolta e sulle azioni in corso: Ministro per la coesione territoriale, 20 February 2014 [on line] Available at: http://www.coesioneterritoriale.gov.it/relazione-sullattivita-svolta-e-sulle-azioni-in-corso/ [Last access: 10 May 2014]

Trotzig, G. (2001). On the (Revised) European Convention on the Protection of the Archaeological Heritage, in *The European Archaeologist*, n.15/2001, Praha: EAA - European Association of Archaeologists, pp. 1-4 [on line] Available at: http://e-a-a.org/tea/archive.htm [Last access: 24 July 2013]

Turok, I. (2009). The Distinctive City: Pitfalls in the Pursuit of Differential Advantage, in *Environment and Planning A*, 41(1), pp. 13-30

UNESCO (1968). Recommendation concerning the Preservation of Cultural Property Endangered by Public or Private works, Records of the General Conference 15th Session, Paris, 1968, Volume 1 – Resolutions, Paris: UNESCO, pp. 139-148 [on line] Available at: http://portal.unesco.org/en/ev.php-URL_

ID=13085&URL_DO=DO_TOPIC&URL_SECTION=201.html [Last access: 24 July 2013]

UNESCO (2011). Recommendation on the Historic Urban Landscape, including a glossary of definitions, Records of the General Conference 36th session, Paris, 25 October – 10 November 2011, Volume 1 – Resolutions, Paris: UNESCO, pp. 50-55 [on line] Available at: http://portal.unesco.org/en/ev.php-URL_ID=48857&URL_DO=DO_TOPIC&URL_SECTION=201.html [Last access: 24 July 2013]

UNESCO World Heritage Centre (2003a). *World Heritage Papers n°7 - Cultural Landscapes: the Challenges of Conservation*, World Heritage 2002 Shared Legacy, Common Responsibility Associated Workshops, Ferrara, 11-12 November 2002, Paris: UNESCO World Heritage Centre [on line] Available at: http://whc.unesco.org/en/series/7/ [Last access: 8 June 2013]

UNESCO World Heritage Centre (2003b). *World Heritage Papers n°9 - Partnerships for World Heritage Cities Culture as a Vector for Sustainable Urban Development*, World Heritage 2002 "Shared Legacy, Common Responsibility Associated Workshops", Urbino – Pesaro 11-12 November 2002, Paris: UNESCO World Heritage Centre [on line] Available at: http://whc.unesco.org/en/series/9/ [Last access: 8 June 2013]

UNESCO World Heritage Centre (2008). *World Heritage Information Kit*. Paris: UNESCO World Heritage Centre. [on line] Available at: http://whc.unesco.org/documents/publi_infokit_en.pdf [Last access: 24 July 2013]

UNESCO World Heritage Centre (2010). *World Heritage Papers n°7 - Managing Historic Cities*. Paris: UNESCO World Heritage Centre [on line] Available at: http://whc.unesco.org/en/series/27/ [Last access: 24 July 2012]

UNESCO World Heritage Centre (2011). *Operational Guidelines for the Implementation of the World Heritage Convention*. Paris: UNESCO World Heritage Centre Available at: http://whc.unesco.org/en/guidelines [Last access: 24 July 2013]

Unioncamere and Istituto Tagliacarne (2009). *Il sistema economico integrato dei beni culturali*. Roma: Istituto Tagliacarne [on line] Available at: http://www.beniculturali.it/mibac/multimedia/MiBAC/documents/1288345729406_rapp_BBCCIntegrato_09.pdf [Last access: 24 July 2013]

Unioncamere and Symbola, eds. (2011). *L'Italia che verrà. Industria culturale: made in Italy e territori*, "I quaderni di Symbola" [on line] Available at: http://www.symbola.net/assets/files/Ricerca%20Industrie%20culturali_1326723510.pdf [Last access: 13 May 2014]

Unioncamere and Symbola, eds. (2012). *L'Italia che verrà: Industria culturale, made in Italy e territori: Rapporto 2012*, "I quaderni di Symbola" [on line] Available at: http://www.symbola.net/assets/fi les/Italia_che_verrà_2012_COMPLETA_1343211887.pdf [Last access: 15 May 2014]

Unioncamere and Symbola, eds. (2013). *Io sono cultura: L'Italia della qualità e della bellezza sfida la crisi: Rapporto 2013*, "I quaderni di Symbola" [on line] Available at: http://www.unioncamere.gov.it/download/2401.html [Last access: 15 May 2014]

United Nations - World Commission on Environment and Development (1987). *Our common future*. New York: United Nations [on line] Available at: http://conspect.nl/pdf/Our_Common_Future-Brundtland_Report_1987.pdf [Last access: 28 October 2014]

United Nations (2008). *Creative Economy Report 2008.The Challenge of Assessing the Creative Economy: Towards Informed Policy Making*. Geneva and New York: United Nations [on line] Available at: http://www.unctad.org/creative-economy [Last access: 30 September 2013]

United Nations (2010). *Creative Economy Report 2010. Creative Economy: A Feasible Development Option*, Geneva and New York: United Nations [on line] Available at: http://www.unctad.org/creative-economy [Last access: 30 September 2013]

United Nations (2013). *Creative Economy Report 2013: Widening local development pathways*. Geneva and New York: United Nations [on line] Available at: http://www.unctad.org/creative-economy [Last access: 30 September 2013]

Usai A. (2013). Creatività e *place-based development*: buone pratiche e strumenti di rigenerazione creativa per la competitività territoriale, in Colavitti, A. M., *Governance* del territorio, Beni Culturali, piano urbanistico. Buone pratiche per la valorizzazione e la competitività, Firenze: Alinea, ISBN 9788860554383

Usai, A. (2013b). Le politiche urbane in Italia e in Europa: dalle sperimentazioni dell'ultimo ventennio alla programmazione 2007-2013, in Colavitti, A. M., *Governance del territorio, Beni Culturali, piano urbanistico. Buone pratiche per la valorizzazione e la competitività*, Firenze: Alinea

Usai A. (2015). Il distretto culturale evoluto: una "rete di reti" per il rinnovamento delle politiche urbane, in *Urbanistica Informazioni*, 263(7), pp. 48-53

Usai A. (2016). *Il distretto culturale evoluto, Beni culturali e pianificazione del territorio nella sfida futura*, Firenze: Altralinea Edizioni

Usai, N. (2010). *Grandi strutture per il tempo libero: Trasformazione urbana e governance territoriale*. Milano: Franco Angeli

Valentino, P.A. (2001). *I distretti culturali: Nuove opportunità di sviluppo del territorio*. Roma: Associazione Civita

Valentino, P.A. (2003). *Le trame del territorio: politiche di sviluppo dei sistemi territoriali e distretti culturali*. Milano: Sperling & Kupfer

van Oers, R. (2010). Managing cities and the historic urban landscape initiative – an introduction, in UNESCO World Heritage Centre. *World Heritage Papers n°7 - Managing Historic Cities*. Paris: UNESCO World Heritage Centre, pp. 7-19 [on line] Available at: http://whc.unesco.org/en/series/27/ [Ultimo accesso 24 July 2012]

Varaldo, R. (2006). Il nuovo modello competitivo e aziendale dei distretti industriali, in *Economia e politica industriale*, n.1/2006, pp. 25-42

Veldpaus, L. and Pereira Roders, A.R. and Colenbrander, B.J. F. (2013). Urban Heritage: Putting the Past into the Future, in *The historic environment*, 4(1), pp. 3-18

Voelpel, D. (2007). Seeds planted to start growing Tacoma's creative class, in *Tacoma News Tribune*, 18 November 2007.

Wacquant, L.J.D. (1992). Introduction, in Bourdieu, P. and Wacquant, L. J., *Réponses pour une anthropologie reflexive,* Paris: Seuil.

Wellman, B. and Hampton, K. (1999). Living Networked On and Offline, in *Contemporary Sociology*, 28(6), pp. 648-654

Wellman, B. and Hampton, K. (2002). The Not So Global Village of Netville, in Wellman, B. and Haythornthwaite, C., eds., *The Internet in Everyday Life*, Oxford: Blackwell, pp. 345-371

Wettenhall, R. (1981). The quango phenomenon, in *Current Affairs Bulletin*, 57(10), pp. 14-22

Williams, R. (1989). *Keywords: A Vocabulary of Culture and Society*. London: Fontana

Williamson, O.E. (1975). *Markets and Hierarchies: Analysis and Antitrust Implications*. New York: The Free Press

Williamson, O.E. (1991). Comparative Economic Organization: the Analysis of Discrete Structural Alternatives, in *Administrative Science Quarterly*, 36(2), pp. 269-296

Wood, S. and Dovey, K. (2015). Creative Multiplicities: Urban Morphologies of Creative Clustering, in *Journal of Urban Design*, 20(1), pp. 52-74

World Commission on Culture and Development (1995). *Our Creative Diversity*. Paris: UNESCO Publishing

Zerman, P.M. (2012). Il federalismo demaniale alla luce del decreto sulla *spending review*, in *Diritto e Pratica Amministrativa*, September 2012, pp. 59-63

Zukin, S. (1995). *The Cultures of Cities*. Oxford: Blackwell Publishers

Zukin, S. (2008). *Naked City: The Death and Life of Authentic Urban Places*. Oxford: Oxford University Press

Zukin, S. (2010). *Consuming Authenticity*, London: Routledge

Cultural Management and Cultural Policy Education

The series exists to foster critical debate and to publish academic research in the field of cultural management and cultural policy as well as to open up a forum for discussions and debate on the topics of cultural management and cultural policy among researchers, educators, scholars, policy makers and cultural managers. It is also intended to provide reference tools for education and lifelong learning on cultural management and cultural policies.

It is becoming more and more evident that education in cultural management and cultural policy cannot and should not be separated from research being conducted in the field. Since its creation, ENCATC has recognised this need and has always been very active in pursuing, publishing, presenting, and disseminating research in cultural management and cultural policy to strengthen the understanding of cultural management and cultural policy issues.

Book Series Editor

ENCATC
European network on cultural management and policy
1, Avenue Maurice
B-1050 Brussels
Belgium

Tel: 0032.2.201.29.12
info@encatc.org; www.encatc.org

Book Series Director

Cristina Ortega Nuere, *3Walks, Spain*

This publication was evaluated by the ENCATC Research Award Jury:

Jury President

Annick Schramme, *ENCATC President, University of Antwerp / Antwerp Management School, Belgium*

Jury Members

Franco Bianchini, *University of Hull, United Kingdom*

Lluís Bonet, *University of Barcelona, Spain*

Jacques Bonniel, *University of Lyon II, France*

Anita Kangas, *University of Jyväskylä, Finland*

Jerry C Y Liu, *National Taiwan University of Arts, Taiwan*

Richard Maloney, *New York University, United States*

Jaime Alberto Ruiz Gutiérrez, *Universidad de los Andes at Bogota, Colombia*

ABOUT ENCATC

ENCATC is the leading European network on cultural management and policy. It is a membership NGO gathering over 100 higher education institutions and cultural organisations in over 40 countries. ENCATC was created in 1992 to encourage the exchange of knowledge, methodologies, experiences, comparative research and regular assessment of the sector's training needs in the broad field of cultural management from a European point of view through a wide range of working groups, projects, activities and events. ENCATC holds the status of an NGO in official partnership with UNESCO, of observer to the Steering Committee for Culture of the Council of Europe, and is co-funded by the Creative Europe programme of the European Union.

ABOUT THE ENCATC RESEARCH AWARD

The ENCATC Research Award on Cultural Policy and Cultural Management was launched in 2014 by the European network on cultural management and policy, ENCATC, to reward the best recently published PhD thesis presented by a young researcher.

This prestigious recognition aims to stimulate academic research in the field of cultural policy and cultural management with an emphasis on its applied implications. The ENCATC Award also has the ambition to contribute to the process of creating a network of scholars who are competent in doing comparative research projects in cultural policy and cultural management.

The ENCATC Research Award is granted each year to a recently published PhD thesis exploring, through comparative and cross-cultural research, issues at stake and taking a step from evaluative (descriptive) to comparative applied research. In addition, this Award is rewarded to a PhD thesis that can inform policymaking and benefit practitioners active in the broad field of culture.

The winning PhD thesis is translated into English or French and published in the Cultural Management and Cultural Policy Education Book Series.

www.peterlang.com